AI for Game Developers: The 2025 Practical Guide
By Brian G. Burton, Ed. D.

ISBN
Digital: 978-1-937336-27-1
Print: 978-1-937336-28-8

Version: 1.10 (Sep, 2025)

Dedication

For Isla, my loving wife. This book is a direct result of your incredible vision and encouragement. Your belief in this project fueled every word, and for that, I am eternally grateful.

Contents

Preface & Introduction

Game development has always been a blend of creativity, problem-solving, and technical skill. Whether you're a solo developer, part of a small indie team, or just starting your journey, you already know how much work goes into turning an idea into something playable. The arrival of powerful, accessible AI tools has the potential to change our workflows in ways that would have seemed impossible just a few years ago.

In just a few short years, artificial intelligence has gone from being an experimental novelty to a daily tool in the game developer's toolkit. What once took months of manual work— blocking out levels, generating dialogue, creating concept art— can now be accelerated, tested, and iterated in days if not hours. The speed of change has been breathtaking, and for many, overwhelming.

When I first began teaching game development over twenty years ago, the workflows we used were almost entirely manual. Students learned by building everything themselves: models, textures, scripts, audio, and marketing materials. That foundation is still important—but the landscape has shifted. Today, there are AI tools that can generate assets, code, and even entire levels. Used well, these tools can help small teams achieve what once required the resources of a full studio with hundreds of developers.

I wrote this book as a guide and reference for my students, past, present, and future, who want to understand how to integrate AI into their development pipelines without losing their creative voice. My goal is to provide a clear, practical introduction to AI-assisted workflows that's grounded in real, accessible tools you can use today.

In making this guide, I made a conscious decision to focus only on AI tools that are in full production—not experimental betas or closed prototypes—and that are capable of creating quality assets ready for game integration. Every tool mentioned here is available to indie developers today, with proven results in real projects.

Artificial intelligence can speed up asset creation, handle tedious tasks, and even suggest creative directions. But it should never dictate your vision. You should still be doing the parts of game development you love most, whether that's designing levels, writing dialogue, composing music, or fine-tuning combat. AI is best used as an assistant, one that amplifies your strengths and steps in to help cover your weaknesses, so you can focus on the heart of your game.

The goal isn't to turn you into a passive user of AI-generated content. It's to make you a skilled, intentional creator who uses AI to work faster, smarter, and more effectively without ever surrendering your creative control. Whether you're here to save time, push boundaries, or learn new ways to bring your ideas to life, your creativity is still the most important tool in your development process. AI is here to give it more horsepower.

This area is developing rapidly. If, while reading this book, you realize there's an AI tool we should have used for a specific step, please let us know. Your feedback will help us improve and ensure that the next edition of this guide remains practical, relevant, and valuable as possible.

Finally, I would be remiss not to disclose that I used AI to help create this book. Specifically for spelling, grammar, chapter images, and improving some of the prompts provided.

Part 1: Pre-Production

Chapter 1 – Ideation & Concept Development

From Blank Page to Playable Idea

Introduction

Every game begins with an idea. Sometimes it's a flash of inspiration while playing another game, sometimes it's a sketch on a napkin, and sometimes, sadly, it's an empty screen staring back at you while your brain refuses to cooperate.
In the indie and small-team world, the idea stage is critical: you don't have years or dozens of designers to sift through ideas. The sooner you can land on a concept worth pursuing, the sooner you can move to building something real.

AI doesn't replace your creativity—it accelerates it. Think of it as a brainstorming partner who never gets tired, doesn't judge your weird ideas, and can instantly generate hundreds of variations for you to refine or cut without guilt.

The Traditional Approach

Traditionally, game ideation might look like this:

- Gather your team (or yourself) in a room.
- Start throwing ideas on a whiteboard or sticky notes.
- Pick a few promising ones, research similar games, and refine.

This method works, but it has limitations:

- **Time-consuming:** Narrowing down takes hours, days, or even weeks.
- **Idea echo chamber:** The same voices and experiences dominate.
- **Limited scope:** You tend to think in familiar genres and mechanics.

AI-Augmented Ideation

AI flips the process by letting you:

- Generate dozens of unique game ideas in minutes.
- Explore combinations that might not immediately come to mind.
- Quickly identify themes or mechanics that might stand out in the market.

- Instant feedback on hairbrained thoughts with a pro and con list.

But, and this is key, **you are still the designer**. AI output is raw material, not a finished design. You'll need to curate, refine, and shape it into something truly yours.

Choosing the Right AI Tools

When it comes to generating ideas, there isn't a single tool that does it all. Each one brings a different strength to the table. Some are designed specifically for game ideation, others are more general AI assistants that can be steered toward creative work, and a few provide visual inspiration that can spark entirely new directions. The tools in this section are here to give you options. Think of them as partners you can call on at different moments in your process. Use them to explore concepts, test variations, and keep your imagination moving forward, but always remember, *you* are the creative engine. The AI is simply here to help you get there faster.

- **ChatGPT / Gemini / Grok / Claude**
  Great for rapid brainstorming and mechanics exploration.
  Best when given constraints (genre, platform, audience).
  https://chat.openai.com
  https://gemini.google.com
  https://grok.x.ai
  https://claude.ai
- **Ludo.ai**
  Specialized for game ideation.

Includes trend analysis, concept scoring, and idea banks.
https://ludo.ai

- **Perchance Generators**
  Quick randomizers for unusual twists (e.g., "Combine racing with underwater exploration").
  https://perchance.org

- **Sudowrite**
  Specializes in generating creative text, including plot twists, character personalities, and dialogue.
  Great for narrative-driven games and fleshing out backstory during ideation.
  https://www.sudowrite.com

- **Artbreeder**
  AI tool for generating and "breeding" images to create unique portraits and landscapes.
  Excellent for quickly producing visual mood boards and character concepts for a pitch.
  https://www.artbreeder.com

Prompt Engineering Principles for Ideation

Ideation with AI is more than just typing a few words and hitting "Enter." To get the most out of your AI brainstorming partner, you need to be intentional with your requests. Think of it as prompt engineering—the art of crafting the perfect prompt to get the best possible result. These principles will help you steer the AI's creativity toward your unique vision, turning it from a simple text generator into a powerful tool for developing your game's core concept.

Be Specific and Provide Constraints: Instead of just a genre and a theme, include a target platform (e.g., "for mobile devices," "for PC with a focus on mouse controls") or a monetization model (e.g., "for a free-to-play game") to get more focused results.

Iterate on AI Outputs: Your first output is a starting point. If the AI gives you a good idea, ask it to refine it even further.

For example, '*Expand on idea #2 by adding a new game mechanic inspired by board games.*'

Use AI as a Creative Mixer: Suggest prompts that deliberately combine disparate ideas to foster unexpected connections.

For example, "*Generate 5 game concepts by combining [Genre 1] with [Gameplay Loop from Genre 2] and a [Narrative Trope].*"

Process: From Prompt to Concept

1. Be specific and set constraints
    Genre, target platform, art style, theme, monetization model, and budget range.

2. Run an initial AI brainstorm
    Gather 10–15 ideas without judgment.

3. Use AI as a Creative Mixer
    Combine different ideas to foster unexpected

connections. E.g., "Generate 5 game concepts by combining [genre 1] with [genre 2] and a [narrative trope].

4. Select the top contenders
   Look for originality, feasibility, and personal excitement.

5. Refine and iterate with focused prompts
   The first suggestion is a starting point. Ask AI to expand mechanics, describe the target audience, and propose visual styles.

6. Document in a Mini Concept Sheet
   Title, core loop, key mechanic, visual direction, and target audience.

Prompt Examples:

- "Give me 5 game concepts combining [genre] with [theme] for [platform or style]."

- "Suggest unique mechanics for a [target platform] game aimed at [audience]."

- "Generate 3 game ideas inspired by [art style] with a focus on [gameplay loop]."

- "Give me 3 tropes associated with [genre], with a list of examples of them done well and badly."

- "Show me how to make [narrative trope] work well with my desired audience without alienating everyone else."

Mini Example: AI-Powered Game Pitch

Prompt:

> Give me 3 original game ideas for a 2D puzzle-platformer that mixes puzzle mechanics with exploration in a cozy-fantasy setting.

Our AI Output (condensed):

1. **Lantern Paths** – You guide magical light orbs through an enchanted forest, solving environmental puzzles to restore balance.

2. **Tea & Treasure** – A traveling tea merchant solves brewing challenges to unlock ancient maps leading to hidden gardens.

3. **Wispwood Chronicles** – A young spirit explores forgotten shrines, arranging runestones to reveal new paths.

Refinement:

- I like #1, but I want more player-driven movement. Prompt AI: *"Add light-based stealth mechanics where shadows are safe zones from wandering spirits."*

Now it's not just puzzle + exploration—it's puzzle + exploration + light-stealth, making it feel unique.

Mini Concept Sheet:

- **Title:** Lantern Paths

- **Core Loop:** Explore → solve light puzzles → unlock new zones → face stealth challenges

- **Key Mechanic:** Light as both a tool and a hazard

- **Visual Direction:** Hand-painted, warm-toned forest environments

- **Target Audience:** Players who enjoy *A Short Hike* and *Ori and the Blind Forest*

Ethical Considerations

- **Originality & Plagiarism:** Run searches to ensure your core concept isn't too close to an existing game. Does the game *feel* too similar to existing Intellectual Property (IP)? AI may suggest names, worlds, or characters similar to existing works.

- **Bias awareness:** AI outputs reflect patterns from its training data—diversify your sources to avoid unintentional bias.

- **Over-reliance and Loss of Creative Voice:** Guard against letting the AI do all the work. If you find yourself mindlessly accepting every idea, you risk losing your unique creative fingerprint. The goal is to use AI to amplify your ideas, not replace them.

- **Ethical Pitfall: The Creative Debt Trap**
  This is the risk of using an AI to generate so many ideas that you lose sight of your own unique creative vision. The trap is to become so reliant on the AI that you spend more time sifting through its output than you do generating original ideas yourself. This can lead to a game that feels like a collection of derivative concepts rather than a truly original work. To avoid this, use AI for inspiration and exploration, but always return to your own core passion to select and refine the ideas that genuinely excite you.

Transition to Next Step

You now have at least one promising game concept documented. Next, we'll expand on it in Chapter 2: The Game Design Document & Narrative Outline, by building the foundation of your Game Design Document (GDD)—the roadmap that keeps your vision clear from prototype to release.

Chapter 2 – The Game Design Document & Narrative Outline

Turning Your Idea into a Roadmap

Introduction

In Chapter 1, we transformed *Lantern Paths* from a rough AI-generated concept to a documented mini-concept sheet. Now, we'll turn that seed into something more substantial: a Game Design Document.

The GDD is the backbone of game development—it's your blueprint, your bible, your communication tool, and your reminder of what you're actually building when the months get long and the scope starts to creep.

Think of the GDD as your game's instruction manual for the team (even if "the team" is just you and your cat). It ensures that the vision you have today is still clear six months from now.

The Traditional Approach

Traditionally, a GDD is a long, sometimes intimidating document created entirely by hand, often in Google Docs, Notion, or Word. It can be dozens of pages outlining:

- Game overview and elevator pitch
- Story and setting
- Gameplay mechanics and controls
- Art and audio direction
- Level design overview
- Technical requirements

While thorough, the traditional approach takes a lot of time upfront and can be overwhelming—especially for solo or small teams.

AI-Augmented GDD Creation

AI changes the game here:

- You can quickly populate first drafts for every section of the GDD.
- AI can **help organize** the structure and keep formatting consistent.
- You can use AI to explore alternative ideas without rewriting everything manually.

The trick is to **never let AI lock in the vision for you**—you're using it to save time, not to make your design decisions for you.

Choosing the Right AI Tools

Building a Game Design Document (GDD) is about clarity and structure. You don't need to make it long—you need to make it usable. The following tools can help you generate text, organize content, and visualize your ideas so the GDD becomes a roadmap you'll actually use. Each of these options brings different strengths, so you can pick the one that best matches your workflow.

Text Generation & Refinement

- **ChatGPT / Claude / Gemini**
  Large language models that can help you draft sections of the GDD, refine descriptions of mechanics or features, and reformat content for clarity.
  https://chat.openai.com
  https://claude.ai
  https://gemini.google.com

Game-Specific Templates

- **Ludo.ai**
  Designed with game developers in mind, Ludo includes built-in templates for pitches, design elements, and even early GDD structures.
  https://ludo.ai

Collaborative Organization

- **Notion AI**
  Integrates directly into Notion workspaces, allowing you to generate and format GDD sections collaboratively. Great for teams who need live, searchable documentation.
  https://www.notion.so/product/ai

- **Milanote**
  A visual-first planning tool. Excellent for organizing the art and audio sections of a GDD, building mood boards, linking reference videos, and mapping out aesthetics.
  https://milanote.com

Long-Form Organization

- **Scrivener**
  Not a generative AI tool, but a powerful writing platform with organizational features that make it easier to maintain long documents like GDDs.
  https://www.literatureandlatte.com/scrivener

Core GDD Sections for Indie Teams

Here's a lean GDD structure that works well for small teams and scales if you grow:

1. Game Overview
   - Elevator pitch (2–3 sentences)
   - Genre, target platform, and audience
   - Unique selling points

2. Story & Setting
   - World description
   - Key characters
   - Story beats

3. Gameplay Mechanics
   - Core loop
   - Player actions and abilities
   - Progression system

4. Art Direction
   - Visual style and influences
   - Color palettes and mood boards

5. Audio Direction
   - Music style
   - Sound effects tone

6. Level Design
   - World structure

- Types of challenges and pacing

7. Technical Details
   - Engine and tools
   - Target performance specs

Prompt Engineering Principles for GDD

Crafting a GDD with AI is an incredible time-saver, but it's not a do-it-and-forget-it process. To turn a good AI output into a great one, you need to think like an engineer, not just a writer. The key is to refine your prompts to guide the AI toward your specific vision, ensuring that the final document is a cohesive, useful blueprint for your game. These principles will help you get beyond the first draft and create a GDD that truly reflects your creative goals.

Use the "Act as" Prompt: Instruct the reader to use prompts that give the AI a persona.

For example, a prompt like, "*Act as an experienced game designer. You are writing a GDD for a cozy-fantasy exploration puzzle game. Your goal is to be concise and inspiring.*"

This can help the AI generate more professional and well-structured content.

Chain Prompts for Better Results: Instead of one long prompt, use a series of smaller, more focused ones.

For example: "*Draft a 1-page story outline for a cozy-fantasy puzzle game where a young spirit must restore light to a forgotten forest."*

Then, *"Based on the story outline, draft a detailed description of the art direction, focusing on the visual feel of the world, lighting, and a few key characters."*

Reverse-Engineer from an Example: If you have access to a well-known game's GDD structure (or even a fan-made one), share it with your AI tool of choice and ask it to adapt the format for their game. This can help the AI produce a more professional and organized document.

Keep Instructions brief but detailed: The majority of AI systems respond better to short, detailed instructions that are easily understood. When coupled with Chained Prompts, this will produce a much better GDD.

Workflow for *Lantern Paths*

Step 1 – Outline the Document
Prompt:

Act as an experienced game designer.

You are writing a GDD for a cozy-fantasy exploration puzzle game.

Your goal is to be concise and inspiring. Create a GDD outline for a cozy-fantasy exploration puzzle game

> titled *Lantern Paths,* where light is both a tool and a hazard.
>
> Include sections for story, gameplay, art, audio, level design, and technical details."

Step 2 – Fill in the Sections with Draft Text from AI
Example output for **Game Overview**:

Lantern Paths is a cozy-fantasy exploration puzzle game for PC and Nintendo Switch.

Players guide magical light orbs through an enchanted forest, using shadows as safe zones to evade wandering spirits.

Combining puzzle-solving, exploration, and stealth, *Lantern Paths* offers a relaxing yet immersive journey for fans of *A Short Hike* and *Ori and the Blind Forest*.

Step 3 – Chain Prompts for Better Results:
Instead of one long prompt, use a series of smaller, more focused ones. For example:

- **Prompt 1:**

> Draft a 1-page story outline for a cozy-fantasy puzzle game where a young spirit must restore light to a forgotten forest.

- **Prompt 2**:

> Based on the story outline, draft a detailed description of the art direction, focusing on the visual feel of the world, lighting, and a few key characters.

- **Prompt 3**:

> Expand on the gameplay mechanics, describing how the player will interact with the light and shadows.

- **Alternatively, Reverse-Engineer from an Example:**
  Give the AI a well-known game's GDD structure (or even a fan-made one) and ask it to adapt the format for their game. This can help the AI produce a more professional and organized document.

Narrative Outline

Your GDD doesn't need a full script yet, but it should contain a **narrative spine**—the key beats that keep the player moving forward.

For *Lantern Paths*, the narrative might be:

1. **Inciting Incident:** The enchanted forest is losing its balance as the Lantern Spirits grow dim.
2. **Early Exploration:** The player learns to manipulate light to restore small sections of the forest.
3. **Midgame Twist:** The wandering spirits are not enemies—they're guardians testing the player.
4. **Climax:** The player restores the final Great Lantern, revealing the forest's true history.
5. **Resolution:** Harmony returns, but hints remain of deeper mysteries.

Prompt example for AI assistance:

> Write a 5-beat story outline for a game where magical light must be guided through a forest, but shadows provide safety from spirits.

Mini Example: GDD Excerpt for *Lantern Paths*

Gameplay Mechanics – Core Loop:

- Explore forest zones to locate Lantern Shrines.
- Solve environmental puzzles by directing light orbs along paths.
- Use shadows to avoid contact with patrolling spirits.
- Unlock new zones and upgrades to your lantern staff.

Art Direction:

- Hand-painted backgrounds with soft, warm lighting.
- Seasonal color shifts to reflect progress (spring → summer → autumn).

Audio Direction:

- Calming, harp-driven soundtrack with ambient forest sounds.
- Spirit encounters marked by low, resonant chimes.

Ethical Considerations

- **Avoid Lifting Storylines or Character Names:** Ensure that AI-generated content is used as a starting point and is refined to avoid direct replication of existing works. Be clear in your documentation which parts were AI-assisted to avoid confusion later.

- **Consistency vs. Collaboration:** If you are a solo-developer working with a freelancer, it's crucial to be transparent about what was AI-generated and what was not. Using an AI-generated art style or narrative as a starting point without full disclosure can feel misleading to human collaborators who expect to be shaping the project from scratch.

- **Ethical Pitfall: The Algorithmic Bias Trap**
  AI models are often trained on GDDs from a narrow range of genres, which can lead to biased output. This pitfall is the risk of the AI inadvertently omitting or deprioritizing elements important to your game's genre, such as emotional depth or a relaxed pace in a cozy game. This can lead to a GDD that feels sterile or incomplete. The designer must be vigilant in identifying and correcting these biases to ensure the GDD accurately reflects the game's unique vision.

Transition to Next Step

With your GDD in place, you've set a strong foundation. Next, in Chapter 3: Narrative, Dialogue & Worldbuilding, we'll expand the *Lantern Paths* worldbuilding, flesh out the characters, and use AI to assist in creating branching dialogue and lore.

Chapter 3 – Narrative, Dialogue & Worldbuilding

Bringing the Forest to Life

Introduction

Now that *Lantern Paths* has a clear concept and a solid Game Design Document, it's time to bring its world to life.
 This stage is about more than just writing a plot—it's about creating a setting, characters, and lore that make the player want to keep exploring.

For indie and small-team developers, this is often where the magic happens, but it's also where scope can spiral out of control. AI can help keep you focused, generate starting material for you to refine, and provide fresh perspectives when you're stuck.

The Traditional Approach

Normally, worldbuilding and narrative design involve:

- Writing setting descriptions and backstory by hand.
- Developing a cast of characters with motivations and arcs.
- Mapping out dialogue for quests, tutorials, and flavor text.
- Creating lore documents to keep everything consistent.

It's rewarding but time-consuming, especially if you are doing it alone.

AI-Augmented Worldbuilding

AI can:

- Suggest character personalities and motivations.
- Generate alternate plot beats for branching storylines.
- Help maintain tone and consistency in dialogue.
- Provide quick drafts for environmental descriptions.

Remember, AI output is a *starting point*. Your unique voice shapes your final result.

Choosing the Right AI Tools

When you are building out your story, characters, and world, different tools shine in different areas. Some focus on narrative structure, others on character voices and interactions, and some on the maps and visuals that make your world feel alive. Here are AI resources you can draw on depending on what part of the process you're working on.

Narrative Tools

- **ChatGPT / Claude / Gemini**
  Large language models that can generate lore, quest ideas, branching dialogue options, and narrative outlines. Great for first-draft material that you can refine.
  https://chat.openai.com
  https://claude.ai
  https://gemini.google.com

- **Sudowrite**
  An AI writing assistant designed for creative writing. Excellent at generating text with flair—whether you need dialogue, character backstories, or a fresh plot twist.
  https://www.sudowrite.com

- **Notion AI / Obsidian**
  Productivity and knowledge-management tools with AI features. Ideal for organizing your lore bible, keeping story notes searchable, and connecting different worldbuilding elements.

https://www.notion.so/product/ai
https://obsidian.md

Dialogue & Character Interaction Tools

- **Inworld AI**
  Creates rich, persistent NPC personalities with
  emotions, memories, and goals. Perfect for characters
  who feel consistent across multiple interactions.
  https://inworld.ai

- **Convai**
  Focuses on real-time, voice-driven NPC conversations.
  Best for immersive player-led dialogue where responses
  need to feel unscripted.
  https://convai.com

- **ElevenLabs.io**
  Industry-leading AI voice generation. Produces natural-
  sounding lines that can be used for prototyping or even
  final in-game delivery.
  https://elevenlabs.io

Worldbuilding Tools

- **Artbreeder**
  A collaborative AI image-generation tool that lets you
  "breed" and evolve portraits, landscapes, and props.
  Excellent for quickly visualizing NPCs, environments, or

cultural styles
https://www.artbreeder.com .

- **Notion AI / Obsidian** *(dual-use)*
  While they shine in narrative organization, these tools are also useful for cataloging maps, geography, and cultural details so you don't lose track of your growing world.
  https://www.notion.so/product/ai
  https://obsidian.md

Tip: Inworld AI is great for complex personalities and emotional depth, while Convai is a strong choice if you want immediate, fluid conversation that feels less scripted. For *Lantern Paths*, we might use Inworld AI for key story characters and Convai for secondary NPCs who offer exploration hints or respond dynamically to the player's actions.

Prompt Engineering Principles for Narrative

While your GDD lays out the core character concept, bringing them to life requires a consistent voice and personality. This is where a strategic approach to prompting becomes invaluable. Instead of simply asking for dialogue, you can use AI to build a rich, reusable "character bible" that ensures every word they say and every action they take feels authentic to their personality. These techniques turn AI from a one-off tool into a creative partner that helps you maintain your vision across the entire game.

- **Establish a "Tone", "Style", and "Voice" First:** Before asking for dialogue or lore, instruct the AI on the narrative tone.

  For example, a prompt like, *"Write a lore snippet for a cozy-fantasy forest, in the style of a wise, ancient storyteller. Use a calm, slightly melancholic tone."*

  This helps prevent generic or bland outputs.

- **Create a Character Profile Prompt:** Suggest a structured prompt for character design. This provides a reusable template.

  For example, a prompt could be, *"Create a character profile for an NPC named [Name]. They are a [Role] and their personality is [Adjective 1] and [Adjective 2]. Their main motivation is [Motivation], but they have a secret flaw: [Flaw]. Give me five sample lines of dialogue."*

  This approach forces the AI to be more detailed and consistent.

- **Prompt for Dialogue Variations:** Encourage the reader to ask for multiple takes on a single line of dialogue to find the best fit.

  For example, *"Give me three ways Bran could say, 'I see you've found a secret path,' with one being playful, one mischievous, and one slightly impressed."*

This helps a developer find the perfect line without having to write it from scratch.

Workflow for *Lantern Paths'* World

Prompt Example – Setting

> Describe a cozy-fantasy forest divided into three main regions, each with unique seasonal colors, creatures, and environmental puzzles.

Refined Output (Condensed)

- **Spring Glades** – Blossoming wildflowers, gentle streams, and puzzles that involve guiding light over water.

- **Summer Canopy** – Dense green leaves casting intricate shadows, puzzles based on shifting dappled light.

- **Autumn Hollow** – Golden leaves, early evening, sunsets, puzzles using long shadow patterns to unlock paths.

Designing Key Characters

Prompt Example – Characters

> Create three main NPCs who help the player restore light to the forest. Include personality, role, and a personal side quest.

Refined Output for *Lantern Paths*

1. **Eira, the Lantern Keeper** – Gentle mentor figure, teaches advanced light control. Side quest: Recover her lost ceremonial lantern.

2. **Bran, the Pathfinder** – Mischievous guide, knows secret routes. Side quest: Clear shadow corruption from his childhood trail.

3. **Soren, the Silent Guardian** – Stoic protector spirit. Side quest: Restore his memory fragments scattered through the forest.

Dialogue and Voice with AI Assistance

AI can quickly generate first-pass dialogue, which you then rewrite for tone and flow.

Example for **Bran, the Pathfinder**:

- **AI Draft:** "Hey, traveler! Fancy seeing someone else out here. The shadows can be tricky—you've got to think like the light."

- **Refined Version:** "Well, well... another wanderer braving the shadows. Keep your eyes on the glow, friend—it knows the way, even when you don't."

For interactive testing, you could:

- Use **Inworld AI** to give Bran a personality profile and scripted memory of meeting the player.

- Use **Convai** to let Bran answer unscripted questions about the forest in real time.

- Use **ElevenLabs.io** to provide Bran's warm, mischievous voice while testing pacing and delivery.

Mini Example: Lore Snippet for *Lantern Paths*

"Long before the first lantern was lit, the forest lived in perfect balance. Day and night danced in harmony, each giving the other room to breathe. But when the Great Lantern dimmed, the shadows grew restless, and the dance faltered..."

Ethical Considerations

- **Originality:** Always run searches to ensure your core concepts aren't too close to an existing work. While AI can create novel ideas, its output is based on existing data and can inadvertently echo storylines or characters from a popular game or book

- **Avoid Accidental IP Use:** Be mindful that an AI might suggest names, worlds, or characters that are similar to existing intellectual property. You should always research names and concepts before they're finalized to avoid legal issues down the road.

- **Bias Awareness:** AI outputs reflect patterns from its training data, which can perpetuate harmful stereotypes. It's your responsibility to review NPC personalities and dialogue for any unintended bias and ensure your characters and story elements are respectful and inclusive.

- **Voice Rights:** If you are using an AI voice tool like ElevenLabs.io, you must ensure you have the legal rights to use that voice model or synthetic likeness. Be transparent with players if a character's voice is AI-generated, especially if it's based on a real person's voice with their permission.

- **NPC Transparency:** When using tools like Convai for real-time, unscripted NPC interactions, make players aware that they are interacting with an AI. This builds trust and sets expectations for the player's experience.

- **Ethical Pitfall: The Empty Room Problem**
  This pitfall is the belief that generating a thousand lines of dialogue will create a compelling character. AI can provide the raw material, but it cannot on its own, create a sense of place or emotional resonance. You must still be the one to carefully curate and place a select few lines that truly matter and to craft a narrative that gives those words meaning.

Transition to Next Step

The story of *Lantern Paths* now has a heartbeat, its setting, characters, and lore are ready to be visualized.

In Chapter 4: Visual Concept Art & Style Guides, we'll use AI to create visual concept art and style guides that bring this world into view before we start building it in-engine.

Chapter 4 – Visual Concept Art & Style Guides

Seeing the World Before Building It

Introduction

With *Lantern Paths*' story, characters, and lore in place, it's time to visualize them.
 Before you ever open your game engine, having a cohesive set of concept art and style guidelines ensures that every visual element—environments, characters, UI—feels like part of the same world.

For indie and small teams, this step is vital. You may not have a dedicated art department, so your concept art and style guide become your "north star" for visual consistency, especially if you work with freelancers or outsource certain assets.

The Traditional Approach

Traditionally, visual concept work involves:

- Hiring a concept artist or working with in-house art staff.
- Creating mood boards from existing art, photos, and media.
- Producing multiple drafts of characters, environments, and props.
- Compiling a style guide that details color palettes, lighting styles, and texture references.

This approach is highly effective but often expensive and time-consuming for small teams.

AI-Augmented Concept Development

AI can:

- Generate quick mood boards from text prompts
- Explore variations on a single concept in minutes
- Produce high-quality placeholder art for prototyping
- Upscale and refine rough drafts to near-production quality

The role of AI here isn't to replace a skilled artist—it's to accelerate exploration and lock in the game's visual identity early.

Choosing the Right AI Tools

When it comes to visual concept art and style guides, the right AI tool can speed up exploration and help you create a consistent look for your game. Some tools specialize in high-quality art styles, others in flexibility or refinement, and a few are designed specifically for game asset workflows. Below are tools you can use to create concept art, mood boards, and style guides that will carry through the rest of development.

Concept Art Generation

- **Midjourney**
  Generates high-quality, stylized concept art. Known for its painterly look and wide adoption in creative industries. Excellent for mood boards and early visual exploration.
  https://www.midjourney.com

- **DALL·E**
  Flexible image generation tool that supports **inpainting** (editing inside an image). Great for iterating on concept art or revising specific details without starting over.
  https://openai.com/dall-e

- **Stable Diffusion**
  An open-source image generation model that gives developers full control over outputs. Highly customizable with community-trained models and plugins.

https://stability.ai/stable-diffusion

- **Krea.ai**
  Real-time AI art generation with a strong focus on iteration and exploration. Useful for quickly trying multiple directions and refining prompts visually.
  https://www.krea.ai

- **Adobe Firefly**
  Adobe's AI art platform, integrated with Photoshop and Illustrator. Excellent for polished concept art, text-to-image exploration, and refining visuals directly within industry-standard creative tools.
  https://www.adobe.com/products/firefly.html

Refinement & Enhancement

- **Magnific AI**
  AI-powered upscaling and detail enhancement tool. Perfect for taking rough concept art or generated images and making them sharper and more polished.
  https://magnific.ai

Game-Specific Asset Support

- **Scenario.gg**
  AI platform designed specifically for game developers.
  Allows you to train the AI on your own art style so the
  generated outputs match your project's unique look.
  https://www.scenario.com

Prompt Engineering Principles for Visuals

Visual concept work with AI is more than just typing a
description; it's about giving the AI a clear, intentional vision to
follow. By treating your prompts as a blueprint, you can guide
the AI to create art that is not just beautiful but also cohesive
and relevant to your game's style. These principles will help you
get beyond generic fantasy art and into the specific visual
identity of your project.

- **Create a Style Prompt:** Advise the reader to create a
  master prompt that defines their game's visual style,
  then reuse it for every image they generate.

  For example, for *Lantern Paths*, this style prompt could
  be: *Hand-painted, cozy-fantasy game, soft lighting, warm
  tones, magical, gentle mist, golden hour.*

  By adding this to the beginning of every prompt, the

developer can ensure that all generated art feels like it belongs in the same world.

- **Use Negative Prompts:** Introduce the concept of "negative prompts," which tell the AI what *not* to include in the image. This is a powerful technique for refinement.

  For example, to avoid overly realistic or generic fantasy art, a negative prompt could be: *photorealistic, gritty, sharp lines, hyper-detailed, unsettling.*

- **Prompt for Specific Angles and Composition:** Include camera angles and composition terms in their prompts to get more useful concept art.

  For example, *an over-the-shoulder shot of a character looking at a glowing shrine or a wide-angle establishing shot of a dense forest with a winding path.*

Mood Boards for *Lantern Paths*

When creating mood boards for *Lantern Paths*, the best place to start is with the three forest regions introduced in Chapter 3: the Spring Glades, the Summer Canopy, and the Autumn Hollow. Each region has its own seasonal palette, mood, and atmosphere, which makes them natural anchors for the game's visual style.

By collecting images, colors, and textures that capture the essence of each space, you create a shared reference point that keeps your art direction consistent. A strong mood board doesn't just inspire—it also serves as a guide for everyone working on the project, from level design to character art.

Prompt Example – Spring Glades Mood Board

> Create concept art of a bright, welcoming forest in early spring. Soft green leaves, blooming

wildflowers, and shallow sparkling streams. The mood should feel fresh, calm, and inviting, with warm sunlight filtering through the trees. Stylized hand-painted look, suitable for a cozy puzzle-adventure game.

Prompt Example – Summer Canopy Mood Board

Generate an illustration of a dense summer forest canopy. Tall trees with deep green leaves, patches of dappled light on the forest floor, and thick vines hanging between branches. The mood should feel vibrant and alive, with strong contrasts between shadow and sunlight. Stylized, painterly textures that emphasize warmth and energy.

Prompt Example – Autumn Hollow Mood Board

Produce concept art of a quiet forest hollow in autumn. Trees covered in golden and crimson leaves, fallen leaves scattered on the ground, and a faint evening glow in the distance. The mood should feel nostalgic and slightly mysterious, with long shadows stretching across the path. Stylized hand-painted style, designed for a cozy puzzle-adventure.

The larger your game becomes, the more important it is to have a unified style guide. AI can help with that as well!

Unified Style Guide Prompt – Lantern Paths

Produce concept art of a quiet forest hollow in autumn. Trees covered in golden and crimson leaves, fallen leaves scattered on the ground, and a faint evening glow in the distance. The mood should feel nostalgic and slightly mysterious, with long shadows stretching across the path. Stylized hand-painted style, designed for a cozy puzzle-adventure.

Use AI to generate multiple variations for each region. Then, refine your favorites and assemble them into a mood board— either in Krita, Photoshop, Canva, or an online whiteboard tool like Miro.

Sample Output from ChatGPT 5:

Lantern Paths - Unified Forest Style

Color Palette

Shadow Green

Soft Glow

Autumn Red

Golden Light

Forest Green

Mood & Themes
- Hand-painted textures
- Cozy exploration
- Balance of light & shadow
- Warm, magical atmosphere
- Seasonal variation (Spring/Summer/Autumn)

Character Visual Development

From Chapter 3, we know our NPCs: Eira, Bran, and Soren. AI can help by generating silhouette explorations, costume variations, and even facial expression studies.

Prompt Example – Character Concept

> Fantasy NPC character design, cozy-fantasy style, named Eira, an elderly Lantern Keeper. She wears layered robes in warm colors, carries a carved wooden staff with a glowing lantern, and has a kind expression. Soft hand-painted aesthetic.

Always iterate. Ask AI to produce **five variations** of Eira, then combine elements you like into the final concept.

46

Advanced Prompting for Visual Consistency

Create a Style Prompt: Create a master prompt that defines their game's visual style, then reuse it for every image generated.

For example, in Lantern Paths, this style prompt could be: *Hand-painted, cozy-fantasy game, soft lighting, warm tones, magical, gentle mist, golden hour.*

By adding this to the beginning of every prompt, the developer can ensure that all generated art feels like it belongs in the same world.

Use Negative Prompts: Negative prompts tell the AI what not to include in the image. This is a powerful technique for refinement.
For example, to avoid overly realistic or generic fantasy art, a negative prompt could be: *photorealistic, gritty, sharp lines, hyper-detailed, unsettling.*

Prompt for Specific Angles and Composition: Include camera angles and composition terms in their prompts to get more useful concept art.

For example, an *over-the-shoulder shot of a character looking at a glowing shrine* or *a wide-angle establishing shot of a dense forest with a winding path.*

Creating the Style Guide

A style guide should include:

1. **Art Style Reference** – Descriptive summary and sample reference or inspirational images.
2. **Color Palette** – Base, accent, and highlight colors for each region.
3. **Lighting Guidelines** – How shadows and highlights are treated.
4. **Character Design Rules** – Proportions, line thickness, and rendering style.
5. **UI/UX Style Notes** – Fonts, icon styles, and menu design guidelines.

For *Lantern Paths*, the style guide might define:

- **Art Style:** Hand-painted, warm tones, visible brush strokes.
- **Lighting:** Soft glow around lanterns, deep but non-threatening shadows.
- **Spring Glades Palette:** Pastel greens, pale yellows, light blues.
- **Summer Canopy Palette:** Rich greens, golden sunlight, muted browns.
- **Autumn Hollow Palette:** Deep oranges, golds, and purples.

Mini Example: Style Guide Excerpt for *Lantern Paths*

Lighting Guidelines:

- Lantern light should be the brightest element in a scene.
- Shadows are tinted with the region's primary and secondary colors.
- Mist effects are semi-transparent with a subtle gradient fade.

Character Rules:

- Friendly NPCs use rounded shapes; antagonistic spirits use angular silhouettes.
- Lanterns and staffs are scaled slightly larger than realistic for visual emphasis.

Ethical Considerations

- **Licensing:** Always verify the license terms for AI-generated images before using them commercially. Some AI tools may have restrictions on commercial use, so it is vital to know your rights before adding assets to your project.

- **Avoiding Imitation:** If AI art is based on an artist's style, ensure you have permission or alter it significantly to avoid imitation. Prompting an AI to generate an image "in the style of" a living artist can be seen as an attempt to devalue a human artist's unique style and may have legal repercussions.

- **Placeholders:** Keep placeholders separate from production-ready assets in your project folder. This prevents confusion and ensures that final assets are original or properly licensed before release.

- **Ethical Pitfall: The Substitution Trap**
  The ethical trap is believing that generating a mood board or a few pieces of concept art is a substitute for a well-defined art direction. You, the developer, still need to do the hard work of creating a clear style guide and ensuring every in-game asset adheres to it, rather than just having a collection of pretty pictures.

Transition to Next Step

Now that we have *Lantern Paths*' art direction and visual identity locked in, it's time to start building, beginning with prototype levels and gameplay systems in Chapter 5: Prototyping, Greyboxing & Full-Level AI.

Part II: Production

Chapter 5 – Prototyping, Greyboxing & Level Design with AI

From Concept Art to a Playable Space

Introduction

With *Lantern Paths*' narrative and visual identity locked in, it's time to get the game playable—fast.

This is where we move from mood boards and story outlines into **interactive spaces**. Our goal here is not perfection but **function**: a working prototype that lets us explore core gameplay, test mechanics, and get feedback early.

For indie and small teams, early prototypes also help secure funding, attract collaborators, or build a community. AI can accelerate this stage dramatically by generating playable layouts, placeholder assets, and even a first-pass lighting setup.

The Traditional Approach

Traditionally, prototyping involves:

- Blocking out basic levels with primitive shapes ("greyboxing").
- Manually creating navigation paths, colliders, and test interactions.
- Iterating until the flow feels right before replacing placeholders with final art.

While effective, this method can take days or weeks just to produce something testable, especially if you're working solo.

AI-Augmented Prototyping

AI can now:

- Generate an entire level layout from a text prompt.
- Create placeholder 3D models for props, obstacles, and buildings.
- Suggest gameplay variations and pacing changes.
- Automatically place lighting, hazards, and checkpoints.

Choosing the Right AI Tools

Prototyping and greyboxing are about speed and iteration. The goal is to get something playable as quickly as possible, then refine it. AI tools can help by generating environments, placeholder assets, and even early mechanics, so you can focus on testing gameplay flow rather than building everything from scratch. Here are some of the most practical tools for indie and small-team developers at this stage.

Prototyping & Playable Generators

During our testing, we found that the available prototyping tools are still in development. Use with caution.

- **Buildbox 4**
  Rapid prototyping and simple AI-assisted level creation. Best for non-coders who want to quickly test gameplay ideas without worrying about scripts. It has its own store for publishing games.
  https://www.buildbox.com

- **Ludo.ai Playable Generator**
  Goes beyond ideation by generating basic playable prototypes. Useful for getting mechanics up and running quickly so you can test early gameplay loops.
  https://ludo.ai

Placeholder Assets & Greyboxing

- **Meshy.ai / Hitem3D.ai**
  Create placeholder props, obstacles, and environment assets. These tools make greyboxes more visually readable while keeping dev time low, helping players and testers understand the flow of a level even before final art is added.
  https://www.meshy.ai
  https://hitem3d.ai

Environment & World Visualization

- **Blockade Labs**
  Generates 360-degree skyboxes or immersive environments directly from text prompts. Ideal for quickly testing the look and feel of a setting before committing to detailed level design.
  https://www.blockadelabs.com

3D Modeling & Rigging Support

- **Masterpiece Studio**
  A suite of AI-powered tools for 3D asset creation, including a generative modeler and rigging solutions. Excellent for building placeholder characters or props that are rig-ready for animation tests.
  https://www.masterpiecestudio.com

Procedural Content Generation

We feel it would be remiss to leave PCG tools from our list. While far from comprehensive, PCG can have the same or better results at creating environments for your game.

- **Speedtree** - for quickly generating detailed foliage and vegetation. https://speedtree.com

- **Superhive** - has a number of PCG tools for Blender. https://superhivemarket.com

- **World Machine** - Procedural terrain generator with erosion and biome simulation. Excellent for natural landscapes. https://www.world-machine.com

- **World Creator** - Real-time terrain generator, faster and more intuitive than World Machine. Great for indies needing rapid iteration. https://www.world-creator.com

- **Fab, Godot,** & **Unity** asset stores include various PCG tools for speeding up your level development. https://www.fab.com/ https://assetstore.unity.com/ https://store-beta.godotengine.org/

New Tools in Beta

- **MythGen.ai** - Mythgen is an AI system for rapid prototyping of entire game levels with physics, assets, and animations. It is compatible with Unity, Unreal, and Godot engines. At the time of this writing, MythGen is in

beta. We are looking forward to using it for development in the near future!
http://MythGen.ai

- **Intangible** - spatial-AI tool for creating interactive 3D scene sketches in the browser with drag-and-drop elements, camera setup, and real-time visualization—great for fleshing out level layouts quickly.
https://www.intangible.ai

Prompt Engineering Principles for Prototyping

Prototyping with AI is all about speed and iteration. By treating the AI as a level designer, you can rapidly generate playable spaces that you can immediately test and refine. The key is to be as specific as possible about the gameplay experience, not just the visual layout, so the AI can produce a truly functional and useful prototype.

- **Specify a "Feel" or "Pacing":** Go beyond describing the physical space and prompt the AI for a specific feel or pace.

  For example, instead of just "a winding forest path," prompt for "*a winding forest path that feels claustrophobic at the start but opens up to a wide, majestic clearing.*"

- **Prompt for Interactivity:** Include gameplay elements and their relationships in their prompts.

58

For example, "*Create a level layout for a puzzle game. A glowing orb must be moved from point A to point B. Include a series of three obstacles that require the player to use a specific mechanic.*"

This forces the AI to think in terms of gameplay, not just aesthetics.

- **Use "If/Then" Logic in Prompts:** This is a powerful technique for getting more complex, usable results.

  For example, "*Create a level layout for a stealth game. If the player is in a lighted area, they are seen by enemies. If they are in a shadow, they are safe. Include at least three distinct paths to the exit.*"

Workflow for *Lantern Paths*

Step 1 – Define Your Prototype Goal
Our first playable *Lantern Paths* level will focus on the light-orb puzzle mechanic and basic stealth interactions with wandering spirits.

Step 2 – Generate the Initial Layout
Prompt to Ludo, Buildbox, or similar tool:

> Create a cozy-fantasy forest environment with winding paths, small glades, and two puzzle areas

> requiring the player to guide light orbs to shrines
> while avoiding patrolling spirits.

AI generates a rough level with paths, open areas, and basic geometry.

Step 3 – Import and Adjust

Bring the generated layout into your engine. Replace or adjust AI-placed props with custom placeholders from Meshy.ai or Hitem3D.ai with your game assets.

Step 4 – Add Gameplay Hooks

- Place interactable shrines.
- Add moving light-orb entities.
- Script patrolling spirit AI with simple vision cones.

Step 5 – Test and Iterate

Run through the level. Ask AI for suggestions on improving flow:

> "Analyze this level layout for pacing and challenge
> curve for a puzzle-adventure game. Suggest 3
> changes."

Mini Example: Greybox of Spring Glades

For *Lantern Paths*, the Spring Glades prototype might include:

- **Starting Area:** Player learns to move light orbs.

- **Puzzle Glade #1:** Light must cross a shallow stream without touching water.

- **Puzzle Glade #2:** Light must be passed between shadows to avoid spirits.

- **Shrine:** Completing both puzzles restores part of the forest's glow.

Prompting for Level Design

To help AI help you, use the following prompting examples to create better maps and concept art. Here we will use some of the prompts that were mentioned earlier in the chapter:

- **Specify a "Feel" or "Pacing":** Go beyond describing the physical space and prompt the AI for a specific feel or pace.

  For example, instead of just "a rope bridge over a river," prompt for "a rickety rope bridge made of logs and vines over a raging river, make it feel unsafe."

- **Prompt for Interactivity:** Include gameplay elements and their relationships in their prompts.
  For example, "Create a level layout for a puzzle game. A glowing orb must be moved from point A to point B. Include a series of three obstacles that require the player to use a specific mechanic." This forces the AI to think in terms of gameplay, not just aesthetics.

- **Use "If/Then" Logic in Prompts:** This is a powerful technique for getting more complex, usable results. For example, "Create a level layout for a stealth game. If the player is in a lighted area, they are seen by enemies. If they are in a shadow, they are safe. Include at least three distinct paths to the exit."

Full-Level AI Possibilities

While AI-generated levels are nice for inspiration and speed, they should be treated as drafts, not final designs. Some tools may overcomplicate layouts or produce unusable geometry. The designer's eye is still essential.

In the future, as level design AI tools mature, we may see production-ready levels generated directly from prompts—but even then, your unique level design sensibilities will remain a key differentiator.

Ethical Considerations

- **Licensing:** Always check licensing for AI-generated geometry before shipping. Some AI tools may have restrictions on commercial use, even for placeholder assets, so it is vital to know your rights before adding anything to your project.

- **Placeholders:** If placeholder assets are made by AI, place them in their own folder structure and replace them with original or licensed versions before final release. This prevents confusion and ensures that final assets are original or properly licensed before your game is shipped.

- **Vision Control:** Avoid letting AI design dictate gameplay—ensure the player experience drives the environment, not the other way around. The AI's output is a draft, and your unique level design sensibilities will remain a key differentiator.

- **Ethical Pitfall: The Unplayable Output Trap**
  The trap is to spend more time fixing the AI's mistakes than it would have taken to build the level from scratch. You, the developer, have a responsibility to your time and budget to use the AI wisely, and this pitfall highlights the danger of blindly trusting the AI to produce a perfect, usable asset.

- **Ethical Pitfall: Black Box Problem**
  This refers to the opaque nature of some AI tools' output. You might receive a perfectly functional greybox level, but without understanding how the AI generated the paths and puzzles, they can't be effectively debugged or iterated on. This is a form of intellectual "debt" that can come back to haunt the project later. Always treat AI output as a draft and be prepared to take it apart and rebuild it by hand if necessary.

Transition to Next Step

With *Lantern Paths*' first playable space complete, it's time to start replacing placeholders with proper assets. In Chapter 6: 2D & 3D Asset Creation, we'll use AI to create the shrines, props, and environmental details that will make the Spring Glades come alive.

Chapter 6 – 2D & 3D Asset Creation

From Placeholder to Polished

Introduction

Our prototype of the Spring Glades level in *Lantern Paths* works mechanically—but it still looks like a collection of gray blocks and placeholder objects. This chapter is about replacing those placeholders with real game assets—environment props, interactable objects, and decorative details—so the world starts to feel alive.

For indie and small teams, AI can be a massive timesaver here. Instead of modeling every object from scratch, we can use AI-assisted tools to create custom, style-consistent assets that match our art direction that was defined in Chapter 4.

The Traditional Approach

Normally, asset creation involves:

- Modeling or asset creation in Blender, Maya, Photoshop, Krita, or similar tools.
- UV unwrapping, texturing, and baking maps by hand.
- Exporting and importing into the game engine.

It works, but for a solo developer, creating even 20–30 custom assets can take weeks.

AI-Augmented Asset Creation

AI tools can now support production-quality asset creation by:

- **Generating base 3D models** from text or reference images, ready for refinement in traditional modeling tools.
- **Auto-texturing and UV unwrapping** models, reducing repetitive setup work.
- **Matching generated outputs to a defined art style**, ensuring consistency across assets.
- **Creating optimized 2D game assets** such as icons, UI elements, decals, and sprites.
- **Producing high-resolution textures** that tile cleanly and support PBR workflows.
- **Accelerating rigging and animation setup** for characters or props with skeletal structures.

- **Scaling assets across variations** (e.g., multiple props, environmental details, or style variants) with minimal extra work.

Choosing the Right AI Tools

When you move from placeholder assets to production-quality models and textures, the right tools can save enormous amounts of time. Some focus on 3D asset pipelines, others on 2D sprites and UI, and a few straddle both worlds. Below are the AI resources well suited for indie and small-team developers at this stage.

3D Asset Creation

- **Meshy.ai**
  Generates 3D models from text or images, with built-in AI-assisted texturing. It can produce props, structures, and game-ready assets quickly.
  https://www.meshy.ai

- **Hitem3D.ai**
  Creates 3D assets directly from descriptions, ensuring models are export-ready for Unity, Unreal, or other game engines.
  https://hitem3d.ai

- **Kaedim**
  Transforms 2D concept art into 3D models. Especially useful for turning character or prop sketches into ready-to-edit geometry.
  https://www.kaedim3d.com

- **Sloyd 2.0**
  A browser-based AI modeling tool with Text-to-3D and Image-to-3D generation. Ideal for indies thanks to presets for different art styles and optimized game-ready exports.
  https://www.sloyd.ai

- **Meshroom**
  An open-source photogrammetry tool that uses AI algorithms to turn photo sets into detailed 3D models with realistic textures. Perfect for capturing real-world objects and converting them into digital assets.
  https://alicevision.org/#meshroom

2D Asset Creation

- **Adobe Firefly**
  Integrated into Adobe's Creative Suite. Great for generating textures, vector graphics, and using generative fill to expand or edit existing images.
  https://firefly.adobe.com/

- **PixelLab**
  An AI art tool specialized for 2D assets—generates

sprites, objects, and icons. Great for pixel-art or
lightweight 2D games.
https://www.pixellab.ai

Hybrid & Experimental Tools

- **Layer**
  An AI tool for generating both 2D and 3D game assets.
  Useful for indies needing flexibility across different
  asset types in one workflow.
  https://layer.ai

Prompt Engineering Principles for Assets

Visual concept work with AI is more than just typing a
description; it's about giving the AI a clear, intentional vision to
follow. By treating your prompts as a blueprint, you can guide
the AI to create art that is not just beautiful but also cohesive
and relevant to your game's style. These principles will help you
get beyond generic fantasy art and into the specific visual
identity of your project. As was previously discussed in Chapter
4, a style prompt and negative prompts are critical for a
consistent style.

- **Create a Style Prompt:** Create a master prompt that
  defines their game's visual style, then reuse it for every
  image they generate.

  For example, for *Lantern Paths*, this style prompt could

be: Hand-painted, cozy-fantasy game, soft lighting, warm tones, magical, gentle mist, golden hour.

- **Use Negative Prompts: Negative prompts** tell the AI what *not* to include in the image. This is a powerful technique for refinement.

  For example, to avoid overly realistic or generic fantasy art, a negative prompt could be: *photorealistic, gritty, sharp lines, hyper-detailed, unsettling.*

- **Prompt for Specific Angles and Composition:** Encourage the reader to include camera angles and composition terms in their prompts to get more useful concept art.

  For example, an *over-the-shoulder shot of a character looking at a glowing shrine or a wide-angle establishing shot of a dense forest with a winding path.*

Workflow for *Lantern Paths*

When creating your assets, remember to begin with an explicit reference to your style guide (Chapter 4) so that your assets maintain their visual harmony.

Step 1 – Identify Needed Assets

From our greybox, we know we need:

- Decorative trees and bushes.
- Lantern shrines (two variants).

- Wooden bridges for puzzle areas.
- Small rock formations and mushrooms.
- Ambient props like fallen leaves, signposts, and lantern poles.

Step 2 – Generate Base Models with Meshy.ai

Prompt Example:

> Generate three variations of a hand-painted fantasy style wooden bridge with a mossy texture, weathered wood, small cracks on the planks, for a cozy-fantasy forest game, sized for a single player to cross, with soft curved railings.

Review the AI outputs, select the best fit, and then export in .fbx or .obj format.

Step 3 – Create Shrines with Kaedim

Upload the shrine sketches from Chapter 4's style guide into Kaedim to get a 3D base model.
Kaedim handles the conversion, and you can adjust proportions in Blender before exporting to your game engine.

Step 4 – Texturing with AI Assistance

Use Meshy.ai's texture generator or Adobe Firefly to create mossy stone textures, wood grains, and leaf patterns that match the Spring Glades palette.

Apply the generated images to the models and adjust for tiling and color balance.

Step 5 – Import and Replace in Engine

Swap out greybox and placeholder objects in Godot, Unity, or Unreal with the new models.
Check collision, scale, and material settings in your engine to ensure that everything matches your earlier design for the level.

Mini Example: Spring Glades Upgrade

Before adding final assets:

- Flat gray blocks for bridges and shrines.
- Default-engine trees and foliage.

After adding final assets:

- Lantern shrines with carved spiral patterns and soft moss texture.

- Wooden bridges with curved rails, slightly worn boards, and green accents.
- Trees with pastel blossoms match Spring Glades' color palette.
- Small glowing mushrooms near puzzle areas for visual guidance.

Tips for Success

- Always reference your style guide when generating an asset. Be sure to add the color palette to your prompt.

- Batch-generate assets that appear together to ensure lighting and texture consistency.

- For *Lantern Paths*, keep edges slightly rounded and avoid overly realistic detail to maintain the hand-painted look.

Ethical Considerations

- **Licensing:** As has been noted previously, always verify the license terms for AI-generated images before using them commercially. Some AI tools may have restrictions on commercial use, so it is vital to know your rights before adding assets to your project.

- **Avoiding Imitation:** If AI art is based on an artist's style, ensure you have permission or alter it significantly to avoid imitation. Prompting an AI to generate an image "in the style of" a living artist can be seen as an attempt to devalue a human artist's unique style and may have legal repercussions.

- **Ethical Pitfall: The Placeholder Trap**
  The trap is to become so accustomed to the placeholder art that it becomes the "final" art, leading to a game that feels generic or legally questionable. AI is a tool for rapid prototyping, not a shortcut for creating final, professional assets without proper licensing or rights.

- **Ethical Pitfall: The Iteration Illusion**
  While AI can generate dozens of iterations quickly, the ethical trap is to believe that quantity equals quality. You might have 50 versions of a tree, but without a clear artistic vision and a critical eye, they are no closer to having the *right* tree for your game. Your human judgment and creative vision are still the most valuable parts of the process. They must be the curator, not just a passive consumer of AI output.

Transition to Next Step

Now that the Spring Glades look the part, it's time to make them move and speak. In Chapter 7: Character Avatars, Animation & Voice, we'll bring Eira, Bran, and Soren into the game with full models, animations, and AI-driven conversations.

Chapter 7 – Character Avatars, Animation & Voice

Breathing Life into NPCs

Introduction

Your characters aren't just shapes moving on the screen— they're the living heart of your game. This chapter is about turning your concept art and character descriptions into fully modeled, animated, and voiced NPCs that connect with players.

For *Lantern Paths*, this means giving Eira, Bran, and Soren not just faces and animations, but personalities players can interact with dynamically.

The Traditional Approach

Traditionally, this stage involves:

- Creating 3D models in tools like Blender, Maya, Character Creator, or MetaHuman.
- Rigging for animation, either manually or with auto-riggers.
- Animating movements by hand or via motion capture.
- Recording voice lines in a studio with actors.

For indie teams, these steps can be expensive and slow, especially if characters require multiple animation cycles and extensive dialogue.

AI-Augmented Character Creation

AI can accelerate every step:

- **Modeling & Rigging:** Generate base models using AI-assisted tools and auto-riggers.
- **Animation:** Procedural animation and motion capture cleanup with AI tools.
- **Voice Acting:** AI-generated voices for prototyping or even final delivery.
- **Real-Time NPC Interaction:** Integrating AI-driven personalities into fully animated characters.

Choosing the Right AI Tools

When it's time to bring characters to life, you'll need tools for creating models, animating them, and giving them voices and personalities. Below are some practical options for indie and small-team developers.

Character Creation

- **Reallusion Character Creator**
  AI-assisted morphing, auto-rigging, and stylized output. Excellent for creating production-ready characters that can be quickly exported to game engines.
  https://www.reallusion.com/character-creator/

- **Unreal Metahuman**
  Generates high-fidelity, fully rigged human characters. Especially powerful for cinematic projects or games requiring realistic avatars.
  https://www.unrealengine.com/en-US/metahuman

- **Ready Player Me**
  A cross-platform avatar system that allows players to create unique custom 3D characters. Saves developers time by outsourcing customization while still integrating seamlessly into games.
  https://readyplayer.me

Animation

- **Cascadeur**
  AI-assisted animation software that helps create realistic physics-based motion. Great for both keyframe animators and non-experts.
  https://cascadeur.com

- **DeepMotion**
  Provides AI-powered motion capture directly from video. Ideal for indie teams without access to a mocap studio.
  https://www.deepmotion.com

- **Radical Motion**
  Delivers real-time body tracking and motion capture solutions. Useful for live prototyping and quick animation workflows.
  https://radicalmotion.com

- **Audio2Face**
  NVIDIA's AI tool for automatic lip-syncing, turning any voice input into matching facial animation.
  https://build.nvidia.com/nvidia/audio2face-3d

- **Uthana**
  A generative AI platform for 3D character animation. Converts text or video into motion data, trained on real mocap datasets for more natural results.
  https://uthana.com

Voice & Interaction

- **ElevenLabs.io**
  Produces natural, emotive AI voice generation. Works well for both prototyping dialogue and final in-game delivery.
  https://elevenlabs.io

- **Inworld AI**
  Creates personality-driven NPCs with emotions, memories, and goals. Adds depth to characters beyond scripted dialogue.
  https://inworld.ai

- **Convai**
  Specializes in real-time, voice-enabled NPC interaction. Perfect for immersive, unscripted conversations that adapt to player input.
  https://convai.com

Prompt Engineering Principles for Character Creation

While your GDD lays out the core character concept, bringing them to life requires a consistent voice and personality. This is where a strategic approach to prompting becomes invaluable. Instead of simply asking for dialogue, you can use AI to build a rich, reusable character bible for you to curate, that ensures every word they say and every action they take feels authentic to their personality.

These techniques turn AI from a one-off tool into a creative partner that helps you maintain your vision across the entire game.

Create a Character Bible Prompt: Create a comprehensive prompt that defines a character's traits, voice, and backstory. This is a critical step before generating dialogue. The prompt should include details like:

[Character's Name] is a [role] with a [personality trait] and a [flaw/wound]. Their backstory is [backstory]. Their voice is [voice description]. Keep the length and pacing to [Length/pacing]

This character bible can be used for all subsequent dialogue and animation prompts to ensure a consistent character across all interactions.

Prompt for Emotional Range: Ask for different emotional states in your prompts.

For example: *Generate three lines of dialogue for Eira, the Lantern Keeper, to say when the player solves a difficult puzzle. One line should sound genuinely surprised, one should be proud, and one should be slightly relieved.*

This helps to create a more nuanced and lifelike character.

Prompt for Physical Actions: Include physical actions in their dialogue prompts.

For example: *Bran says, 'Keep your eyes on the glow, friend,' while gesturing with a mischievous smile.*

This can be directly handed to an animator or used as a reference for a procedural animation system.

Workflow for *Lantern Paths* - Integrating Convai

In Chapter 3, we first mentioned using Convai for Bran, the Pathfinder, to provide spontaneous exploration hints. Now, we'll integrate it into the full character workflow:

1. **Model & Rig the Character**
   - Create Bran in Reallusion Character Creator, ReadyPlayerMe, or Unreal Metahuman, using the art style guide from Chapter 4.

2. **Connect to Convai**
   - Use Convai's SDK for Unreal/Unity integration to link Bran's model to Convai's conversational AI.

3. **Lip-Sync & Animation**
   - Follow Convai's live dialogue tutorial for accurate lip movements based on the model that you used..

   - Add additional idle animations and gestures using Reallusion iClone, Cascadeur or DeepMotion.

4. **Gameplay Hooks**

- Allow Bran to answer questions dynamically ("Where's the next shrine?") or comment on player progress.

- Ensure that there are fallback scripted lines for when AI connection is unavailable.

Mini Example: Bran in Action

- Player: *"How do I get to the Autumn Hollow?"*

- Bran (via Convai): *"Ah, the Hollow's tricky. Keep to the shadow of the old oak until you see three lanterns in a row, then follow the glow."*

- Lip-sync is generated in real-time, and Bran gestures toward the correct path.

Ethical Considerations

- **Be transparent with players when an NPC is AI-driven.** While AI-powered characters can enhance immersion and make the game world feel more alive, it's important to be upfront with your players. Some players may be sensitive to the use of AI, and this transparency can build trust and set clear expectations for the experience. The goal is to avoid the "imposter AI" trap where players feel they've been deceived into believing an NPC is more intelligent than it actually is.

- **Avoid collecting unnecessary player voice data.** Player voice data can be considered personal information, so you must be mindful of privacy regulations like the Children's Online Privacy Protection Act (COPPA) in the United States and the General Data Protection Regulation (GDPR) in the European Union. These regulations require you to limit data collection, obtain verifiable parental consent for players under a certain age, and provide players with the right to access, correct, or delete their data. Companies have faced millions in fines for failing to comply with these regulations.

- **Test extensively to prevent inappropriate AI responses.** AI models can be unpredictable, and in-game dialogue systems can sometimes generate responses that are toxic, offensive, or break from the established tone and lore of your game. To mitigate this, you should employ a rigorous testing process to identify and fix potential issues before the game goes live. This is an ongoing process that requires continuous monitoring and updates. You should always test when the AI system receives an update or patch in case there is a change that breaks your characters personality..

- **Ethical Pitfall: The Voice Appropriation Trap** There is the danger of creating a voice model that sounds too much like a real, well-known actor or public figure. This can lead to legal issues and alienate players who are sensitive to the unauthorized use of a person's identity. It's an important reminder to use AI voices with care and respect for the human element they are based on.

- **Ethical Pitfall: The Uncanny Valley of Personality**
  When an AI character sounds or acts almost human but has small, unsettling inconsistencies that break immersion. This is a form of the "uncanny valley" applied to personality and can be just as jarring as a poorly rigged character model. It reinforces the need for extensive human testing and refinement of AI character behavior.

Transition to Next Step

With *Lantern Paths'* characters now visually, vocally, and interactively complete, the next stage is to immerse them in a soundscape that matches their world. In Chapter 8: Audio: Music, Sound Design & Voice Acting, we'll craft the forest's ambient sound, mystical effects, and memorable musical themes.

Chapter 8 – Audio: Music, Sound Design & Voice Acting

Giving the Forest Its Voice

Introduction

Players don't just *see* your game, they *hear* it. In *Lantern Paths*, the forest's gentle ambience, the chime of lanterns, and the whispers of guardians will shape how players feel as they explore. Audio can set mood, guide gameplay, and deepen immersion as much as visuals or story.

For indie and small teams, audio is often left for last or outsourced entirely, but AI tools now make it possible to create custom, high-quality audio assets without a dedicated composer or studio.

The Traditional Approach

Traditionally, this stage involves:

- Composing music in a DAW (Digital Audio Workstation) or hiring a composer.
- Recording Foley (real-world sounds) for SFX.
- Hiring voice actors and booking studio sessions.
- Editing, mastering, and integrating audio into the engine.

It works beautifully—but for a small team, it can be slow, costly, and limit iteration.

AI-Augmented Audio Production

AI can now:

- Compose original, royalty-free music from prompts.
- Design sound effects from text or reference audio.
- Generate placeholder or even final voice lines.
- Automate audio mastering and loop creation.

Choosing the Right AI Tools

Music, sound, and voice are essential for creating an immersive player experience. AI tools can now compose music, organize samples, generate voice acting, and even separate elements from existing audio tracks for remixing or reuse. Here are some of the most practical tools available for indie and small-team developers.

Music Composition

- **Suno**
  High-quality AI-generated music with the ability to add
  custom lyrics and generate in multiple genres. Great for
  both background tracks and featured songs.
  https://suno.ai

- **Aiva**
  Specializes in composing orchestral, ambient, or
  thematic music. Excellent for adaptive soundtracks that
  match the mood of gameplay.
  https://www.aiva.ai

- **Boomy**
  A fast, simple music generation platform. Best for quick
  iterations, placeholder tracks, or small indie projects
  needing rapid audio prototypes.
  https://boomy.com

Sound Design

- **Sononym**
  AI-assisted sound organization and sample discovery.
  Helps build and manage an SFX library by finding
  relationships between sounds.
  https://www.sononym.net

- **LALAL.AI**
  Not a generative tool but an AI-powered deconstruction platform. Splits audio tracks into vocals, instruments, and stems—useful for isolating sounds, creating loops, or repurposing music.
  https://www.lalal.ai

- **AudioStrip**
  A quick, browser-based resource for isolating music from dialogue or separating effects from background audio.
  https://audiostrip.co.uk

Voice

We already looked at character voice as part of performance in Chapter 7. Here in Chapter 8, we are focusing on voice as audio design and how it fits into the overall soundscape alongside music and effects.

- **ElevenLabs.io**
  Provides realistic, emotive AI voice synthesis, music, and SFX. Excellent for prototyping or final delivery.
  https://elevenlabs.io

- **Inworld AI**
  Creates NPC dialogue systems with persistent personalities and adjustable emotional tones. Ideal for games with interactive storytelling.
  https://inworld.ai

Prompt Engineering Principles for Audio

Your game's soundtrack and sound design are just as important as its visuals in creating a memorable experience. Instead of simply generating a song or a single sound effect, you can use AI to build a rich, layered soundscape that matches the emotional tone of your game. By using specific and detailed prompts, you can turn a generic soundscape into a dynamic audio experience that guides the player and deepens their immersion in the world. This approach makes AI a creative collaborator, not just a sound-making machine.

Specify Mood and Emotion: Go beyond genre and describe the emotional core of the sound.

For example, instead of just "calming music," prompt for "*a short, looping background track for a cozy-fantasy forest that evokes a sense of gentle melancholy and hopeful exploration.*"

This guides the AI to produce a more nuanced and emotionally resonant result.

Prompt for Audio Layers: Use prompts to build a soundscape from individual layers, rather than one single prompt for the entire scene.

For example: "*Create a 30-second loop of quiet forest ambience with birdsong and a gentle breeze.*"
Then, in a separate prompt, "*Generate the sound of a small stream trickling over mossy stones.*"

This allows for more precise control and better mixing in the game engine.

Use Audio References in Prompts: Use well-known game audio references in their prompts.

For example, "*Generate a musical theme for a hub area that feels like the music from A Short Hike, with a gentle, acoustic feel.*"

This provides the AI with a clear, established style to reference. Note that some AI tools do not permit specific references to artists or specific songs.

Workflow for *Lantern Paths* – Spring Glades Audio

Sound is one of the most powerful ways to shape atmosphere, yet it's often one of the last areas small teams focus on. By integrating AI early, you can quickly generate placeholder audio, experiment with different moods, and then refine toward production-quality soundscapes. The following workflow shows how to apply AI tools to create music, sound effects, and voice in *Lantern Paths* while keeping the process efficient and iterative.

Step 1 – Define the Audio Atmosphere
Our Spring Glades should feel peaceful but alive:

- Ambient forest with birds, wind in leaves, and soft streams.
- Musical theme in a major key with gentle harp and strings.

- Chimes when the player interacts with lanterns.
- Soft hum around patrolling spirits.

Step 2 – Compose the Region Theme with Aiva or ElevenLabs Music

Prompt Example:

> *"Create a 2-minute looping background track for a cozy fantasy forest. Gentle harp and soft strings, light percussion, slow tempo, calming mood."*

Export the track and set it to loop seamlessly in the engine.

Step 3 – Generate SFX

- Use **Sononym** to organize existing forest and magical sounds.
- Create missing sounds (e.g., glowing lantern activation) via Foley or AI generation.
- Layer subtle environmental noises so no two areas sound exactly the same.

Step 4 – Voice Acting with ElevenLabs or InWorld

Generate Eira's guiding lines with a warm, grandmotherly tone; Bran's with playful energy; and Soren's with a deep, steady cadence.
Test in-game with Convai for unscripted exchanges and Inworld AI for persistent character memories.

Step 5 – Integrate into Engine

- Assign audio triggers to specific zones.
- Use 3D positional audio for environmental SFX (e.g., waterfalls grow louder as you approach).
- Ensure dialogue volume balances with music and effects.

Mini Example: Audio Layer for Puzzle Shrine

Audio layers create immersion by blending music, sound effects, and environmental cues. In *Lantern Paths*, shrine interactions are a perfect example of how layered audio design builds atmosphere.

Scenario

When the player approaches a shrine:

- A soft **wind** fades in, signaling a transition into a sacred space.

- The shrine emits a **low hum**, tuned to harmonize with the existing background music.

- When the lantern's light activates, a **crystalline chime** rings out, marking the moment of success.

- Spirits in the distance add **echoing whispers**, enhancing tension without overwhelming the scene.

AI Prompt Example

"Generate a soft ambient wind loop at 48 kHz, 24-bit, 10 seconds long, with no sudden changes in volume."

"Create a sustained, low-frequency hum tuned to C minor, designed to blend seamlessly with background music."

"Produce a crystalline chime effect that lasts 2 seconds, bright but not piercing, for puzzle completion."

"Generate distant, whispering vocal textures with heavy reverb for an eerie but subtle background layer."

By layering these elements, the shrine encounter becomes more than just a visual interaction. Each audio cue reinforces the emotional tone—mystery, tension, and release—without requiring additional dialogue or cutscenes. AI tools make it possible to generate each element quickly, refine them with consistent settings, and blend them into a polished soundscape.

Tips for Success

- **Match industry standards for audio quality.**
  Always set AI tools to output at 48 kHz, 24-bit, the standard for game-ready audio. This avoids problems when exporting into engines like Unity or Unreal.

- **Keep your acoustic space consistent.**
  If your soundscape relies on reverb, echo, or other spatial effects, make sure the settings match across all generated audio. Inconsistent reverb tails can be more jarring than missing audio altogether.

- **Save raw and processed files.**
  Store the original AI output alongside edited or normalized versions. This allows you to reprocess sounds later without losing the original quality.

- **Think in layers, not single sounds.**
  Complex moments like the shrine activation or boss encounters work best when built from multiple audio tracks (ambient + effects + musical cues) that blend dynamically.

- **Balance novelty with restraint.**
  AI tools can generate endless variations of sound, but too much experimentation can clutter your mix. Define the emotional goal first, then generate just enough to achieve it.

Ethical Considerations

- **Licensing:** As with all types of assets, always verify licensing for AI-generated music and SFX before shipping. Be aware that some AI tools may have restrictions on commercial use, so you must know your rights before adding anything to your project.

- **Voice Transparency:** Be transparent with voice actors if AI is used for temporary or final lines. The ethical choice is to inform them about your workflow and to ensure you have a clear contract that outlines their consent, usage rights, and fair compensation if their voice is used to train an AI model.

- **Ethical Pitfall: Identity Misappropriation**
  This refers to the pitfall of generating voices that are based on an amalgamation of real people's voices without their consent. Even if not a direct clone of a specific actor, an AI voice could inadvertently replicate the unique tonal qualities of a human voice without permission or compensation.

- **Ethical Pitfall: The Sonic Mismatch Trap**
  The trap is believing that a collection of high-quality, but stylistically mismatched, AI-generated sounds can create a cohesive soundscape. A sound that works in a "cinematic trailer" might feel jarring and out of place in a cozy, quiet forest. The audio designer remains the curator of the final experience and must ensure that all audio assets work together in harmony.

Transition to Next Step

With the Spring Glades now looking and sounding alive, the next focus is Programming & Gameplay Systems in Chapter 9, where we'll make *Lantern Paths'* light-orb puzzles, stealth mechanics, and progression loop fully functional with AI-assisted coding.

Chapter 9 – Programming & Gameplay Systems

Bringing Mechanics to Life

Introduction

So far, we have designed *Lantern Paths'* world, built the environments, and filled them with characters, sound, and style. Now it's time to make the game playable in full, programming the light-orb puzzles, stealth mechanics, and progression systems that make the experience engaging.

For indie developers, coding can be the steepest hurdle, especially if the team is light on technical expertise. This is where AI-assisted coding, visual scripting, and emerging systems like Vibe coding can dramatically shorten the gap between design and execution.

The Traditional Approach

Traditionally, gameplay systems are coded manually in languages like C# (Unity), C++ (Unreal), Visual Scripting, or GDScript (Godot).

- Developers write scripts for player movement, AI behaviors, triggers, and UI.
- Systems are tested and debugged iteratively.
- Complexity grows quickly, increasing the risk of bugs and rework.

This method can be slow, especially for designers who are more comfortable with creative work than raw programming.

AI-Augmented Programming

AI can help by:

- Generating first-draft scripts from natural language prompts.
- Suggesting optimizations and fixing bugs.
- Explaining how specific code works in plain language.

Choosing the Right AI Tools

Programming and gameplay systems can be one of the most time-intensive parts of development. AI coding assistants and visual tools can help you write, debug, and test scripts faster, whether you're building core mechanics, prototyping features, or balancing game systems. Below are some of the most practical options for indie and small-team developers.

Code Assistance & Generation

- **GitHub Copilot**
  Provides real-time code suggestions and completions inside your editor. Speeds up repetitive coding tasks and reduces syntax errors.
  https://github.com/features/copilot

- **Replit Ghostwriter**
  A browser-based AI coding companion. Great for teams working online or developers who prefer a lightweight environment without a full IDE.
  https://replit.com/ghostwriter

- **Cursor**
  An AI-powered code editor designed for rapid iteration. Integrates AI deeply into the workflow, making it easier to experiment with gameplay systems.
  https://cursor.com

- **Ludus Engine**
  Generates Unreal Engine scripts or Blueprints directly from natural language prompts, best for quickly prototyping mechanics without extensive coding knowledge.

https://ludusengine.com/

- **Claude (by Anthropic)**
  A powerful LLM capable of generating, explaining, and debugging code. Its extended context window makes it particularly effective for working with large scripts or complex systems.
  https://claude.ai

Testing & Reliability

- **Ponicode**
  Uses AI to automatically write unit tests. For small teams, this is invaluable for ensuring code robustness and reducing bugs without manual test-writing.
  https://www.ponicode.com

Game Systems & Balancing

- **Machinations**
  Analyzes and balances game economies, progression systems, and loops using AI and simulation. Can test thousands of playthroughs to ensure difficulty and pacing feel right.
  https://machinations.io

Tools to Watch – Vibe Coding

These tools are still evolving but represent the next wave of "intent-driven" development, where you describe the *feel* of a mechanic rather than writing code line by line. While not yet as production-ready as Copilot or Cursor, they're worth keeping an eye on:

- **Qodo (formerly CodiumAI)**
  Supports intent-based coding with AI-driven code reviews and automatic test generation. Aims to make "vibe prompts" more reliable by validating results as you iterate.
  https://www.qodo.ai

- **Tabnine**
  An autocomplete-driven AI assistant that learns your coding style. It's moving toward intent-based assistance by adapting outputs to project-specific patterns.
  https://www.tabnine.com

- **Eve Programming Environment**
  A research-driven environment that interprets declarative, intent-focused instructions. Still experimental, but influential in shaping how vibe coding concepts evolve.
  https://witheve.com

- **Aider**
  An AI pair-programming tool designed for natural language prompts. Works directly from your terminal, making vibe coding possible in lightweight workflows.
  https://aider.chat

Prompt Engineering Principles for Programming

Your game's core gameplay loop is the engine that drives the experience, and AI can help you build that engine more efficiently. Instead of relying on a single, massive prompt, you can use a series of focused prompts to generate modular, reusable code that fits your game's unique design. This approach turns AI into a coding partner, helping you develop the program as you build your game.

- **Prompt for Modularity:** Prompt the AI to generate modular, reusable code.

  For example, instead of asking for a single script for the *Spring Glades* puzzle, prompt the AI to create a *"Puzzle Manager" that can be used for all future puzzles*.

  This not only saves time but also makes the code base more organized and easier to debug later.

- **Prompt with Specific Parameters:** Include specific, adjustable parameters in their prompts.

  For example, *"Create a script for a wandering enemy AI. Include a variable for patrol_speed and detection_radius so I can easily tweak its behavior in the editor."*

  This makes the AI-generated code far more useful and ready for in-engine implementation.

- **Prompt for Code Explanations:** When the AI generates a block of code, immediately follow up with a prompt asking for an explanation of what the code does, in plain language.

For example, *"Explain this C# script line-by-line in simple terms, focusing on how it makes the orb move."*

This turns the AI into a mentor, helping the developer learn to code as they build their game.

Vibe Coding – Programming by Intuition

Vibe coding is a newer approach to programming where the developer describes the "feel" or intention of the system in natural language instead of writing detailed code line by line. Instead of focusing on syntax, you tell the AI what the gameplay should *feel like*.

For example, *"make the player's jump feel light and springy, with just enough hang time to clear small gaps."*

The AI then translates that description into code, adjusting variables, physics values, or scripts to match the intended *vibe*. Developers can refine the result iteratively, giving more feedback like *"make it a little snappier"* or *"add more weight on landing."*

For small teams, vibe coding lowers the barrier to entry and allows designers to focus on player experience rather than raw technical details. It doesn't replace traditional coding, but it gives non-programmers a more intuitive way to shape mechanics through *intent-driven prompts*.

Example – Light-Orb Puzzle in *Lantern Paths*
Prompt:

> *"Make the orb move slowly along a set path, glowing brighter when near a shrine, and dimmer in shadow. If touched by a spirit, reset to starting position."*

AI generates the base code for movement, light intensity adjustments, and collision detection.
You then "vibe" tweak by saying:

> *"Make the glow transition smoother, and add a pulsing animation when near the shrine."*

This is especially powerful for non-coders because it turns programming into an **interactive design dialogue**.

Visual Scripting – Building Gameplay Without Code

Visual scripting systems like Unreal's Blueprints, Unity's Visual Scripting, and Godot's VisualScript let you build gameplay logic with nodes and connections instead of writing code.

- Easier for designers and artists to experiment.
- AI can still assist—by suggesting node structures or generating example graphs.

Example – Spirit Patrol Behavior in Blueprint:

1. Create patrol points as variables.
2. Use a loop to move the spirit between points.

3. Add a detection radius node—if player enters, trigger chase.

With AI assistance, you can get a suggested Blueprint flowchart or node list based on your prompt.

Workflow for *Lantern Paths*

AI can help streamline programming, but the key is to structure your workflow so that each step builds on the last. Here's how the core gameplay systems for *Lantern Paths* come together using AI-assisted tools.

Step 1 – Define Core Systems

Before you write a single line of code, outline the systems that will carry the gameplay. For *Lantern Paths*, the three pillars are:

- **Light-orb puzzle logic** – Players must guide or position magical light to solve challenges.

- **Spirit AI patrol and detection** – Non-player entities guard certain areas and react to player presence.

- **Shrine activation and progression save system** – Each shrine unlocked records player progress and advances the story.

By defining these upfront, you ensure that your programming is always tied back to core mechanics rather than drifting into features that don't serve the game.

Step 2 – Choose Your Build Approach

Different projects call for different coding strategies. *Lantern Paths* can be developed in several ways:

- **AI-assisted scripting** – Use AI to scaffold C# scripts in Unity or C++ code in Unreal, then refine them manually.

- **Visual scripting** – Leverage node-based systems for faster iteration, especially when testing puzzle mechanics.

- **Vibe coding** – Explore "feel-first" adjustments, using natural language prompts like "make the orb glow warmer" to fine-tune gameplay sensations early.

Choosing the right approach depends on your comfort level and how quickly you want to get a playable loop running.

Step 3 – Implement With AI Assistance

Now it's time to put the tools to work. Each one serves a role in the coding process:

- **Ludus** – Generate Unreal Engine scripts for puzzles and shrine activation systems, saving time on repetitive logic.

- **GitHub Copilot** – Refine complex systems like Spirit AI patrol patterns and detection logic, filling in boilerplate and suggesting optimizations.

- **Cursor** – Debug and optimize scripts in real time, allowing for rapid iteration and immediate testing.

Together, these tools help you move from outline to functioning gameplay quickly, while still keeping code readable and maintainable.

Step 4 – Test in Greybox & Replace Assets

Don't polish too early. Begin with placeholder objects to confirm that your systems work as intended. Once your light-orb puzzles, Spirit AI, and shrine saves are reliable, swap in the production-ready assets you created in Chapter 6. This ensures that your visuals and gameplay come together without wasted effort on redoing art for broken systems.

Mini Example: Shrine Activation Logic

One of the core systems in *Lantern Paths* is shrine activation. When a player guides a charged LightOrb into the shrine's activation area, the shrine should respond with animation, audio, and progression logic.

Base Pseudocode (before AI assistance)

```
if (LightOrb enters ShrineArea && OrbIsCharged) {
    PlayActivationAnimation();
    PlaySound("ShrineChime");
    UnlockNextArea();
}
```

This simple version gets the job done, but it lacks polish. The shrine activates abruptly, and the progression system doesn't wait for the player to *see and feel* the moment.

AI Prompt (refinement request)

Add a particle effect when the shrine activates and delay the area unlock until the activation animation finishes.

AI-Refined Pseudocode

```
if (LightOrb enters ShrineArea && OrbIsCharged) {
    PlayActivationAnimation();
    PlaySound("ShrineChime");
    SpawnParticleEffect("LightBurst", ShrineCenter);
```

```
// Wait until the animation completes
OnAnimationComplete("ActivationAnimation", () => {
  UnlockNextArea();
  SaveProgress("ShrineUnlocked");
});
}
```

What Changed and Why

- **Particle effect added** → Creates a visual burst of light for feedback.

- **Delayed progression unlock** → Ensures players experience the animation and chime before the area opens, heightening impact.

- **Save system included** → Tracks progress so the shrine stays unlocked even if the game is reloaded.

Tips for Success

- **Let AI handle the repetitive work.**
  Use AI to generate boilerplate functions, repetitive logic, and debugging helpers. This keeps you focused on creative problem-solving rather than tedious setup code.

- **Keep core gameplay logic under human review.**
  Systems that define your game's identity, like puzzle

109

rules, AI behavior, or progression, should always be double-checked and refined by you. AI can draft, but you make sure it feels right and remains stable.

- **Mixing AI-generated code and visual scripting? Document everything.**
  If you're blending node-based systems with AI-assisted scripts, take the time to explain how they connect. Good documentation avoids confusion later and makes collaboration easier.

- **Iterate quickly, but test thoroughly.**
  AI can speed up implementation, but rapid iteration can also hide bugs. Test greybox builds frequently to confirm that mechanics work before you layer in assets.

- **Use vibe coding for feel, then refine with structure.**
  Early on, prompts like *"make the lantern glow warmer"* can get you close to the right experience. Later, lock that behavior into explicit values and functions so your systems remain predictable.

- **Treat AI outputs as drafts, not finished code.**
  Even good AI-assisted scripts may include inefficiencies, security concerns, or unintended complexity. Use them as a starting point, then optimize and simplify.

Ethical Considerations

- **Verify AI-generated code doesn't contain unlicensed or copied code from other projects.** Large Language Models (LLMs) are trained on vast amounts of data, which can include open-source code, copyrighted code, and proprietary code. While these tools are great for generating boilerplate or simple functions, they can sometimes reproduce entire code snippets from their training data. Without checking, you could inadvertently ship code that violates a license or infringes on another developer's copyright. _Always_ run a _license check_ and review the code to ensure its originality.

- **Keep player data handling secure—AI tools may not understand full privacy requirements.** AI models are not inherently trained on data privacy laws like GDPR or COPPA. If you use an AI to generate a script for handling player data, it may not include the necessary security protocols or compliance measures. You are the one responsible for ensuring that all player data, especially sensitive information such as personal details or voice data, is collected, stored, and processed securely and in accordance with all relevant regulations.

- **Best Practice: Maintain Readable, Documented Code.** AI-generated code is often functional but not always clean or well-documented. To ensure your project can be debugged, scaled, or shared with future collaborators, treat AI output as a first draft. Take the time to refactor, add comments, and format consistently so that the code remains readable and maintainable.

This extra step avoids frustration down the road and shows respect for anyone who might work on your project after you

- **Ethical Pitfall: The De-Skilling Trap**
  This refers to the risk of a developer becoming overly reliant on AI to the point where they lose their foundational programming skills. While AI can handle boilerplate or repetitive logic, a developer still needs to understand the core principles of game architecture and design to create a stable, fun game. AI is meant to *augment* their skills, not replace the core knowledge that makes you a great developer.

- **Ethical Pitfall: The Legacy Debt Trap**
  The danger of overreliance on AI is creating a patchwork of opaque, auto-generated code that no human can understand. This "Frankenstein" approach leads to *legacy debt*: scripts so tangled and undocumented that they cripple future development. The ethical risk is passing along an unmanageable codebase that others can't reasonably work with. To avoid this trap, always combine AI assistance with human oversight and documentation.

Transition to Next Step

Now that *Lantern Paths'* gameplay systems are functional, we can focus on Chapter 10: QA & Playtesting, where AI will help us find bugs, balance gameplay, and understand how players experience the forest.

Part III Testing, Launch & LiveOps

Chapter 10 – QA & Playtesting

Making the Game Fun and Flawless

Introduction

With *Lantern Paths*' gameplay systems up and running, it's time to make sure the game works as intended and delivers the experience we've been designing for.

Quality Assurance (QA) and playtesting aren't just about finding bugs—they're about validating fun, improving player flow, and ensuring the game feels right from start to finish.

For indie and small teams, this stage often receives less attention than it deserves, usually because of time or budget constraints. AI tools now allow us to run more frequent, thorough tests without the need for a large QA team.

The Traditional Approach

Traditionally, QA and playtesting involve:

- Manual testing for bugs, glitches, and performance issues.
- Recruiting playtesters to try the game and give feedback.
- Collecting feedback through surveys, interviews, or observation.
- Manually analyzing bug reports and balancing data.

Effective, but slow, resource-heavy, and limited by the number of testers you can find or afford.

AI-Augmented QA & Playtesting

AI can now:

- Run automated play sessions to test navigation, collisions, and triggers.
- Generate heatmaps showing where players spend time or get stuck.
- Detect performance bottlenecks and suggest optimizations.
- Analyze playtester chat logs or feedback for recurring issues.
- Simulate thousands of runs to test balance and pacing.

Choosing the Right AI Tools

Testing and analytics are critical for ensuring that your game is stable, fun, and balanced. AI tools can automate repetitive QA tasks, simulate thousands of playthroughs, and uncover insights in player data that would be hard to detect manually. Here are some useful options for indie and small-team developers.

Automated QA & Playtesting

- **nunu.ai**
  An AI-powered platform for QA automation. Focuses on streamlining bug detection and reducing manual testing workloads.
  https://nunu.ai

- **modl.ai**
  Provides automated playtesting bots, QA reporting, and game balancing. Especially useful for small teams that can't run large-scale in-house QA.
  https://modl.ai

Analytics & Player Insights

- **GameAnalytics AI**
  Analyzes player behavior to predict retention, measure engagement, and highlight problem areas. Helps teams understand how players actually interact with their

117

game.
https://gameanalytics.com

Prompt Engineering Principles for QA

Playtesting and QA are no longer just about squashing bugs; they're about gathering data and improving the overall player experience. AI can't tell you if your game is fun, but it can help you get to a fun experience faster. By treating AI as a data-analysis partner, you can turn a mountain of bug reports and player feedback into actionable insights. These prompting techniques will help you structure your requests so you can automate the tedious parts of QA and focus on making your game feel right.

Prompt for Test Scenarios: Prompt the AI to generate a list of test scenarios for a specific feature or level.

For example: *"Create a list of 10 test scenarios for the Spring Glades level, focusing on the light-orb puzzles and stealth mechanics. Include scenarios for both success and failure."*

Prompt for Bug Reporting: Use a structured prompt to generate a detailed bug report from a simple description.

For example, *"A player reported that the light orb disappeared after touching a spirit. Write a detailed bug report that includes the bug name, a clear description, steps to reproduce, and the expected vs. actual behavior."*

Prompt for Feedback Synthesis: Use AI to analyze large

amounts of qualitative feedback.

For example: *"Analyze the following 50 playtester comments. Identify the top three most common pain points and suggest actionable solutions for each."*

Workflow for *Lantern Paths*

AI tools don't just help you build the game—they also help you test it, balance it, and prepare it for players. Quality assurance and analytics are often overlooked by small teams, but skipping them leads to bugs, broken progression, and frustrated players. Here's a practical workflow for applying AI-assisted testing and analytics in *Lantern Paths*:

Step 1 – Define What to Test

Identify the systems that need to be reliable before you launch:

- Puzzle mechanics (light-orb activation, shrine progression).
- AI behaviors (Spirit patrol routes, detection logic).
- Save/load system (shrines remain unlocked, progress is tracked).
- Performance targets (smooth framerate on intended hardware).

Having a checklist ensures your QA tools are focused where it matters most.

Step 2 – Automate Playtesting

Use AI playtesting bots to simulate how different players might interact with your game:

- **modl.ai** → Run automated playthroughs to check puzzles can't be bypassed and that shrines activate as intended.

- **nunu.ai** → Automate repetitive QA tasks, like regression testing after code changes.

This reduces manual testing time and exposes hidden problems early.

Step 3 – Monitor Gameplay & Balance

Track how systems behave at scale and adjust the game balance:

- **GameAnalytics AI** → Analyze puzzle completion times, drop-off points, and overall player engagement.

These tools help you tune the game experience, not just fix bugs.

Step 4 – Validate With Human Review

AI testing is powerful, but it can't replace actual players. After automated passes:

- Run targeted playtests with real people to check if puzzles feel satisfying and fair.

- Compare AI data vs. human feedback to identify potential problem areas.

Step 5 – Iteration

Prioritize fixes based on AI and human feedback combined. Retest changes with AI bots to confirm stability before live playtests.

Mini Example: AI Playtest Scenario for Spring Glades

Automated testing isn't just about finding bugs—it also reveals design flaws and balance issues. In *Lantern Paths*, AI bots can simulate hundreds of playthroughs in the Spring Glades region, showing where puzzles and stealth elements succeed or fail.

Bot Objective

Simulate a player attempting to:

- Reach both shrines in the Spring Glades.
- Solve the light-orb puzzles at each shrine.

121

- Avoid detection by patrolling spirits along the path.

Playtest Results

- **98% success rate** at the first shrine: Both AI bots and human playtesters completed the light-orb puzzle without major issues.

- **62% success rate** at the second shrine: A narrow choke point caused repeated detection by spirits. Bots flagged this as a high-failure zone, and human testers reported the same frustration.

Adjustment Proposed by AI Analysis

- **Solution:** Add an alternate shadow path around the choke point.

- **Reasoning:** Keeps tension in place but introduces a second option for players who struggle. The goal isn't to make the shrine "easy," but to ensure progression doesn't stall due to a single punishing bottleneck.

AI playtesting doesn't just replicate human behavior—it highlights where *players and bots fail in the same spot*. This overlap is a strong indicator that the issue is systemic, not individual skill. By adjusting the level design, you preserve challenge while improving fairness and flow.

Tips for Successful AI QA

- **Balance AI with human playtesting.**
  AI bots are excellent at stress-testing systems, but they can't measure *fun*. Always validate AI findings with real players to catch issues like pacing, clarity, or satisfaction that bots can't evaluate.

- **Use AI heatmaps for design insight, not just bug hunting.**
  Many AI QA tools provide heatmaps showing where bots fail, cluster, or get stuck. Treat these as level design feedback—opportunities to adjust flow, add alternate paths, or rebalance encounters—not just as "fix this bug" alerts.

- **Test early, test often.**
  Don't wait until your game is "almost finished" to start QA. Use small, frequent builds with AI bots to find and fix issues as you go. This prevents late-stage crunch and avoids building polish on top of unstable systems.

- **Correlate AI data with player progression.**
  Combine analytics (like drop-off points or puzzle completion rates) with AI playtest data. If both bots and humans fail at the same step, it's a strong sign the issue is systemic, not skill-based.

- **Document adjustments after each test cycle.**
  AI can generate a flood of data. Keep a changelog of fixes and balance tweaks tied to each test pass so you can measure progress and avoid reintroducing old bugs.

Ethical Considerations

AI-assisted QA can dramatically speed up your workflow, but it also introduces new ethical and technical risks. The responsibility to use these tools wisely falls squarely on you, the developer.

- **Don't collect personally identifiable information from testers without consent.** Player privacy is paramount. Personally Identifiable Information (PII) can include a name, address, or even a voice recording. Before you collect any PII, you must provide a clear and transparent explanation of what data you'll be collecting, what it will be used for, and how you'll protect it. You also need to obtain explicit consent from your testers. A best practice is to have testers use "dummy" or fake information for things like checkout processes or sign-ups to protect their privacy. It's also important to be aware of and comply with regulations like the GDPR and COPPA.

- **If using AI bots on multiplayer systems, avoid interfering with live player experiences.** While AI bots can be used for things like load testing or finding exploits, their presence in a live multiplayer environment can feel unfair or disruptive to human players. A bot might be designed to exploit a game mechanic in a way that a human wouldn't, which can create a frustrating and confusing experience. If you must use bots in a live game, it's a best practice to clearly inform players when they're interacting with a bot or to only use them in a closed or controlled environment.

- **Clearly label AI-generated bug reports to differentiate from human tester feedback.** AI-powered bug reporting tools can analyze data and generate reports that can look a lot like a report written by a human tester. However, AI reports lack the human insight and context that a human tester can provide. By clearly labeling which reports were AI-generated and which were human, you can avoid confusion and ensure that the right kind of feedback gets to the right person on your team.

- **Ethical Pitfall: The Dehumanization of QA**
  This refers to the pitfall of replacing human testers with AI bots without considering the impact on the industry. While AI can handle tedious and repetitive tasks, a human QA tester is a valuable, creative partner who can provide unique insights and nuanced feedback that a bot cannot. AI should be used to augment and empower human QA, not to replace it.

- **Ethical Pitfall: The False Sense of Security Trap**
  AI won't catch "fun factor" issues. The trap is to believe that because an AI bot has successfully navigated a

level, the level is "good." AI bots can find technical bugs, but they can't tell you if a puzzle is intuitive or if the pacing is right.

Transition to Next Step

With bugs reduced and gameplay balanced, it's time to share *Lantern Paths* with the world.

In Chapter 11: Marketing & Community Engagement, we'll use AI to craft marketing materials, build community hype, and set up store pages that convert curious players into fans.

Chapter 11 – Marketing & Community Engagement

Telling the World About Your Game

Introduction

Making a great game is only half the journey. The other half is making sure people know it exists. For indie and small teams, marketing can feel overwhelming, especially without a dedicated PR department.

AI tools can act as your *creative marketing assistant*, helping you brainstorm campaigns, create promotional assets, and maintain a consistent brand voice across platforms. For *Lantern Paths*, our goal is to communicate its cozy-fantasy world, puzzle-adventure gameplay, and emotional storytelling in a way that draws in our target audience, players who enjoy *A Short Hike*, *Ori and the Blind Forest*, and *Spiritfarer*.

The Traditional Approach

Traditionally, indie game marketing involves:

- Writing press releases and blog posts.
- Creating screenshots, trailers, and key art.
- Posting regularly on social media.
- Reaching out to journalists, streamers, and influencers.

This works, but it's time-intensive and often competes with development work.

AI-Augmented Marketing

AI can now:

- Generate key art, thumbnails, and promotional banners.
- Write store page descriptions optimized for search and engagement.
- Create trailer scripts and even edit video footage.
- Suggest social media content calendars.
- Analyze community engagement trends and recommend timing/content changes.

Choosing the Right AI Tools

Marketing and community engagement can be just as critical as making the game itself. AI tools can help you generate promotional art, create teaser videos, write store descriptions, and analyze how players respond to your marketing efforts. Here are some practical options for indie and small-team developers.

Creative Assets

- **Midjourney / DALL·E / Stable Diffusion / Adobe Firefly**
  Generate promotional art, key visuals, and concept images for use in trailers, store pages, and social media.
  https://www.midjourney.com
  https://openai.com/dall-e
  https://stability.ai/stable-diffusion
  https://firefly.adobe.com

- **Runway / Gemini VEO3**
  Provides AI-assisted video editing, motion graphics, and generative media. Useful for quickly producing trailers or polishing marketing videos.
  https://runwayml.com
  https://gemini.google.com/

- **Pika Labs**
  A text-to-video platform designed for teasers, short animations, and promotional shorts. Ideal for creating quick, attention-grabbing content without a full animation pipeline.
  https://pika.art

Copywriting & Engagement

- **ChatGPT / Gemini**
  Excellent for drafting store descriptions, press releases, blog content, and social posts. Can also refine tone and style for different audiences.
  https://chat.openai.com
  https://gemini.google.com

- **Synthesia**
  Generates realistic AI video avatars. Can be used to create professional announcements, marketing clips, or even in-game tutorials presented by a virtual host.
  https://www.synthesia.io

Analytics

- **Helicone**
  Tracks backend engagement with AI-generated content, helping you measure what resonates with your audience.
  https://www.helicone.ai

- **GameAnalytics AI**
  Monitors player interest and retention, even from early builds or demos, to help you gauge market response and refine your outreach strategy.
  https://www.gameanalytics.com

Prompt Engineering Principles for Marketing Content

Game marketing isn't just about showing off your game's features; it's about connecting with your audience on an emotional level. AI can help you craft compelling, targeted content that resonates with players, but it requires more than a simple request. By using advanced prompting techniques, you can turn a generic marketing prompt into a strategic tool that generates content optimized for emotional hooks, platform-specific tones, and even A/B testing, making your marketing machine as smart as your game.

Use the "Act As" Prompt: Instruct the AI to act as a specific persona to generate more targeted content.

For example, *act as a social media strategist for a cozy-fantasy game.*

Draft five tweet ideas for our launch week, focusing on engagement and viral potential.

Prompt for Emotional Hooks: Have AI to generate content that focuses on the emotional experience of the game, rather than just the features.

For example, instead of "Our game has puzzles," prompt for, *Write a social media post that evokes the feeling of a peaceful, late-night puzzle session in our game.*

Use the "A/B Testing" Prompt: Use AI to generate multiple versions of a piece of content for A/B testing.

For example, *Generate two versions of a headline for our game's Steam page. One should focus on the puzzle-solving challenge, and the other should focus on the cozy, relaxing atmosphere.*

Workflow for *Lantern Paths* Marketing

Marketing doesn't start at launch—it begins as soon as you have something worth sharing. AI tools can help you shape your message, create polished content, and keep your community engaged without overwhelming your time or budget. Here's how the marketing pipeline for *Lantern Paths* comes together step by step.

Step 1 – Define the Core Messaging

Before you design trailers or post on social media, clarify what your game is about in a way that resonates with players.

AI Prompt:

> *Summarize Lantern Paths in 3 sentences for a store description. Cozy-fantasy puzzle adventure, hand-painted visuals, light-orb mechanics, stealth, and narrative depth.*

Example Output:
Lantern Paths is a cozy-fantasy puzzle adventure where you guide magical light through enchanted forests, solving environmental challenges and evading wandering spirits. Explore vibrant seasonal regions, restore the Great Lantern, and uncover the forest's forgotten history. Hand-painted visuals and a gentle soundtrack create a world you'll want to explore at your own pace.

This message becomes the backbone of your marketing—it informs your trailer script, your social media voice, and your store listings.

Step 2 – Create a Launch Trailer with AI Help

A strong trailer is the single most important marketing asset you'll produce. AI tools can streamline the process:

- **Script Writing** → Use ChatGPT or Claude to draft a 60-second narrative script tailored to your audience.

- **Visual Assets** → Capture in-engine gameplay and supplement with generated stills or shorts for atmospheric transitions.

- **Editing** → Use Runway or Pika Labs to cut footage, add motion text, and color grade for consistency.

- **Music** → Generate original soundtrack snippets with ElevenLabs, Aiva, or Suno, ensuring the tone matches the cozy-fantasy vibe.

AI doesn't replace creative vision here—it accelerates production and helps you hit a professional quality bar.

Step 3 – Build a Social Media Calendar

Consistency matters more than volume. A clear posting schedule ensures your audience knows when to expect new content.

AI Prompt:

> *Create a 3-month content plan for an indie puzzle-adventure game with cozy-fantasy visuals, targeting Twitter, Instagram, and TikTok.*

Example Schedule:

- **Mondays** → Devlog GIF or short clip (showcase a new mechanic).

- **Wednesdays** → NPC spotlight posts with quotes (pair AI-generated images with ElevenLabs.io voice clips).

- **Fridays** → Puzzle of the Week teaser to spark discussion and anticipation.

This gives you a repeatable rhythm while leaving space for spontaneous updates.

Step 4 – Community Engagement

Marketing doesn't stop with broadcasting—it's about conversation. Use AI to keep your community spaces active and positive:

- **AI-assisted Discord bot** → Trained on lore and devlog content, it can answer common questions and keep fans engaged.

- **AI moderation tools** → Flag toxic behavior early to maintain a welcoming environment.

- **Interactive content** → Share AI-generated "behind the scenes" concept variations and let the community vote on their favorites.

This approach makes your players feel like collaborators, not just customers
.

Mini Example: Social Post for Bran, the Pathfinder

Image: AI-generated portrait of Bran with a warm forest background.
 Text:

> "Meet Bran, the Pathfinder. He knows every shadow and shortcut in the forest—but will he share them with you?
> #LanternPaths #IndieGameDev #CozyGames"

Ethical Considerations

AI can be a powerful marketing assistant, but it also introduces new ethical and technical risks that can affect your reputation and your relationship with your community.

- **Always disclose if promotional visuals are AI-generated, especially if they differ from in-game graphics.** Transparency is a fundamental principle of ethical AI use. AI images can create unrealistic expectations for players, leading to disappointment and a loss of trust when the final game doesn't match the promotional art. A simple disclaimer, like *"This content was generated with the assistance of AI,"* can build credibility and reassure your audience.

- **Use AI moderation tools carefully—don't rely solely on automation for community safety.** AI can be a powerful tool for moderating online communities by flagging toxic content and removing spam. However, AI models can be biased or make mistakes, potentially censoring or misinterpreting a legitimate player's comment.

137

Relying solely on automation for community safety can lead to a dehumanized environment and can sometimes disproportionately affect marginalized groups. It's a best practice to use AI as an assistant to human moderators, not a replacement.

- **Ethical Pitfall: The Community Deception Trap.** This refers to the risk of using AI to generate fake community engagement, such as automated social media comments, positive reviews, or bot-driven forum posts. This can quickly erode player trust and is a major ethical violation that can lead to platform bans. Community building must be a genuine, human-driven process, and any artificial engagement can and will be seen as a form of manipulation.

- **Ethical Pitfall: The Community Deception Trap** This refers to the risk of using AI to generate fake community engagement, such as automated social media comments, positive reviews, or bot-driven forum posts. This can quickly erode player trust and is a major ethical violation that can lead to platform bans.

- **Ethical Pitfall: The Hype Without Substance Trap** This is a crucial ethical pitfall in game marketing. The trap is to use AI to generate stunning, but unrepresentative, marketing materials that over-promise and under-deliver on the actual gameplay experience. This can lead to negative reviews, player churn, and a damaged reputation.

Transition to Next Step

With your marketing machine running and your audience growing, it's time to think beyond launch.
 In Chapter 12: Post-Launch & LiveOps, we'll explore how AI can help you update *Lantern Paths*, run seasonal events, and keep players engaged long after release.

Chapter 12 – Post-Launch & LiveOps

Keeping the Forest Alive After Release

Introduction

Launching your game is not the end of your project; it's the start of a new phase. For modern indie titles, post-launch updates and live operations (LiveOps) can extend a game's lifespan, keep the community engaged, and open opportunities for new revenue streams.

In *Lantern Paths*, our story-driven campaign is complete at launch, but we can keep players returning through seasonal events, quality-of-life updates, and community-driven content. AI can help design, manage, and monitor these updates without overwhelming a small team.

The Traditional Approach

Traditionally, LiveOps for small teams involves:

- Planning content updates months in advance.
- Manually creating new levels, quests, and assets.
- Monitoring player feedback via forums and social media.
- Releasing bug-fix patches and balance changes.

It works, but it's often reactive, slow, and can pull resources away from starting the next project.

AI-Augmented Post-Launch

AI can:

- Propose seasonal event concepts and narrative hooks.
- Generate event-specific assets quickly.
- Analyze player behavior to suggest balance changes.
- Automate patch note generation.
- Predict churn rates and suggest retention strategies.

Choosing the Right AI Tools

LiveOps and ongoing game support require more than just creativity; they demand consistency, monitoring, and efficient workflows. AI tools can help small teams manage analytics, automate repetitive tasks, and generate fresh content for events, all while keeping systems stable and costs under control. Below are some of the some useful options for sustaining your game after launch.

Monitoring & Analytics

- **Helicone**
  Monitors AI-driven systems, API usage, performance, and costs. Invaluable for small teams managing budgets and ensuring AI integrations are stable.
  https://helicone.ai

- **GameAnalytics AI**
  Tracks player behavior, retention, and engagement over time. Useful for measuring the impact of seasonal events or content updates on player activity.
  https://gameanalytics.com

- **Datadog**
  Not a generative AI tool, but a powerful performance monitoring platform that uses AI/ML for error tracking. It can automatically detect and alert you to server issues, crashes, or performance bottlenecks.
  https://www.datadoghq.com

Content Creation & Updates

- **Ludo.ai**
  Generates new quest ideas and mini-game concepts to keep content pipelines fresh.

https://ludo.ai

- **Meshy.ai / Hitem3D.ai**
  Create event-specific props and decorations quickly, allowing for themed seasonal content without large art-team overhead.
  https://www.meshy.ai
  https://hitem3d.ai

- **Aiva / Suno / Eleven Labs**
  Compose original music tracks for seasonal events, trailers, or special in-game experiences.
  https://www.aiva.ai
  https://suno.com
  https://elevenlabs.io

Automation & Workflow Support

- **Adept**
  Automates complex tasks such as adding new items to an in-game store, scheduling seasonal events, or even generating and posting social media content to keep your community engaged.
  https://www.adept.ai

Prompt Engineering Principles for Ongoing Engagement

Launching your game is a huge accomplishment, but it's just the first step in a long-term relationship with your players. Keeping a community engaged and a game feeling fresh months or even years after release can be a full-time job for a small team. By using AI as your LiveOps assistant, you can automate the brainstorming process and generate new content ideas based on real player data, allowing you to stay focused on the creative work that keeps your players coming back for more.

Prompt for Community-Driven Content: Use AI to brainstorm content ideas based on player feedback and community trends.

For example, "*Analyze player comments on our Discord and suggest three new cosmetic items for our in-game store that align with the most requested themes or ideas.*"

Prompt for Retention Strategies: Use AI to propose solutions for common LiveOps problems.

For example, "*Our player retention is dropping after the first week. Suggest three new in-game goals or rewards that could encourage players to return for the second week.*"

Prompt for A/B Testing Event Details: Use AI to generate multiple versions of an event's details to see what performs best.

For example, *"Generate two versions of a social media post for our Lantern Festival event. One should emphasize the limited-time rewards, and the other should focus on the new story content."*

Workflow for *Lantern Paths* Post-Launch

Step 1 – Plan Seasonal Events with AI Help

Prompt:

> Suggest three cozy-fantasy seasonal events for *Lantern Paths*, each with a light-based theme, new puzzles, and unique NPC dialogue.

Example Output:

1. **Lantern Festival** – The forest glows with floating lanterns; players collect rare light fragments for festival lanterns.

2. **Shadow's Rest** – A quiet, twilight event with puzzles based on extending shadows.

3. **Blooming Light** – Spring event where flowers open when lit, unlocking secret paths.

Step 2 – Design the Lantern Festival Event

Narrative Hook
Once a year, the Lantern Keeper lights hundreds of floating lanterns to honor the Great Lantern's renewal. Players help gather rare light fragments scattered throughout the forest.

Event Gameplay

- Timed quests: Find fragments before lanterns rise.

- New puzzles: Redirect light across the river using festival lanterns.

- Cosmetics: Festival robes for the player character.

AI-Assisted Asset Creation

- **Meshy.ai** for decorative lantern models.

- **Aiva** for a festival soundtrack with soft percussion and string harmonies.

- **Inworld AI / Convai** for new seasonal dialogue from Eira and Bran.

Step 3 – Monitor Player Engagement

- Use **GameAnalytics AI** to track participation rates.

- Adjust quest timers or difficulty based on early event data.

- Helicone can monitor real-time AI interactions to ensure NPCs perform well.

Step 4 – Automate Patch Notes

AI Prompt:

> Write patch notes for a cozy-fantasy game update, adding a Lantern Festival event, festival robes cosmetic, and two new light puzzles.

Example Output:
Lantern Festival arrives in the forest! Help Eira and Bran gather rare light fragments before the lanterns rise. New seasonal puzzles await, along with limited-time festival robes for your character.

Mini Example: Seasonal Lantern Festival Event

LiveOps shines when you can add meaningful seasonal content without overwhelming your team. AI tools help streamline the process, from concept to deployment, making it easier to keep your players engaged with fresh experiences.

148

Event Concept

The "Lantern Festival" marks the start of autumn in *Lantern Paths*. Players gather light-orbs, decorate shrines, and unlock limited-time rewards during the festival week.

AI-Assisted Workflow

- **Ludo.ai** → Generates side-quest ideas such as "collect rare glowing herbs to craft festival lanterns."

- **Meshy.ai / Hitem3D.ai** → Create props like festival lanterns, banners, and autumn decorations in a fraction of the time.

- **Aiva / Suno** → Compose a new background track for the event with soft strings and wind chimes, giving it a warm, celebratory tone.

- **Adept** → Automate adding event items to the in-game store and scheduling limited-time challenges.

- **Helicone / Datadog** → Monitor server performance during peak player activity and flag any unusual spikes in errors or lag.

- **GameAnalytics AI** → Track participation rates and measure how many players return specifically for the festival.

149

Tips for Successful AI-Driven LiveOps

- Keep event durations short (1–2 weeks) to maintain urgency.

- Reuse base mechanics with small twists to reduce dev time.

- Use AI to create themed variations of existing assets.

- Maintain a public event roadmap to build anticipation.

Ethical Considerations

- **Don't place main story content behind temporary events.** In a live service game, you can use AI to design new seasonal events, quests, and mini-games to keep players engaged. However, it is an important ethical line to never lock core story content behind these temporary events. A player should be able to experience the complete narrative of your game at their own pace, and not be forced to log in during a specific time period to access key plot points or missions.

- **Be transparent about event rewards and availability.** Players should always know what they are playing for and for how long. If you use AI to create live events, you must be transparent about the rewards and the event's duration. This helps players plan their time and ensures that they don't feel manipulated or misled.

- **Avoid aggressive retention tactics that frustrate players.** AI analytics can provide powerful insights into player behavior, including when they are likely to stop playing or "churn". However, this data can be used to create overly aggressive retention tactics, like a sudden increase in difficulty to encourage in-app purchases or a one-time-only offer that pressures a player to buy.

- **Ethical Pitfall: The Algorithmic Gating Trap**
  This refers to the pitfall of using AI analytics to create content or monetization strategies that unfairly target specific players. For example, an AI might learn that a player is likely to quit and then offer them a highly discounted item to keep them playing, or it might create a "grind" that is specifically designed to frustrate them into buying an in-game item. This is a highly unethical use of AI that can harm players and lead to a toxic community.

- **Ethical Pitfall: The Unsupervised AI Trap**
  The trap is to deploy an AI system for LiveOps, such as a dynamic NPC or an automated quest generator, and then is not properly monitored, allowing it to produce inappropriate content or create an experience that deviates from the game's core vision.

Transition to Next Step

With *Lantern Paths'* live content plan in place, we're ready to wrap up the development pipeline by looking at Chapter 13: AI Ethics in Game Development, ensuring that every AI-powered choice we've made is responsible, transparent, and fair to players and creators.

Chapter 13 – AI Ethics in Game Development

Responsible Innovation for Small Teams

Introduction

AI can supercharge game development—but it can also introduce risks, both to your team and to your players. For indie and small studios, ethics isn't just a legal checkbox, it's part of building trust, protecting your creative integrity, and ensuring your game doesn't unintentionally harm players or other creators.

In *Lantern Paths*, we have shown how to use AI in almost every phase of development. This chapter is where we step back and ask: *Did we use these tools responsibly—and did we keep our creative heart intact?*

Why AI Ethics Matters

Unethical AI use can:

- Damage your studio's reputation.
- Result in legal action or platform bans.
- Alienate your player base and collaborators.

On the other hand, responsible AI use can:

- Build long-term trust.
- Position you as a forward-thinking developer.
- Avoid costly rework from last-minute compliance issues.

Core Ethical Principles for AI in Game Development

As AI becomes woven into the game development pipeline, developers need clear principles to guide its use. These principles aren't meant to limit creativity; they serve as a foundation for building responsibly, ensuring that AI supports rather than undermines the work of human creators. For indie teams and small studios, where resources are tight and trust with players is crucial, following ethical guidelines helps prevent long-term problems such as unmanageable codebases, biased content, or exploitative design.

In this section, we'll outline the core ethical principles that can help you use AI tools wisely, protect your creative vision, and maintain fairness for your players and collaborators.

1. Transparency

- Disclose when significant AI-generated content is used, especially if it differs from the in-game art style.

- For *Lantern Paths*, marketing posts, and promotional art created with Generative AI are tagged as "AI-assisted" to avoid misleading players.

2. Consent

- Only use the training data you have the right to.

- If you use voice cloning (e.g., ElevenLabs.io), ensure the voice actor has given explicit permission.

- For NPC dialogue in *Lantern Paths*, we created original character backstories rather than training on someone else's copyrighted text.

3. Fair Compensation

- If an AI tool is replacing a role, consider whether human collaborators should still be compensated for their creative direction.

- Example: AI-generated music for *Lantern Paths* was composed with ElevenLabs, but a human composer should review and adjust the arrangements.

4. Avoid Harmful Bias

- AI models can reflect biases from their training data.

- Check for stereotypes in AI-generated NPC dialogue or quest prompts.

- During development, we reviewed Inworld and Convai scripts to ensure representation was respectful and not stereotypical.

5. Data Privacy

- Never collect or store player data without consent.

- AI analytics tools like GameAnalytics should be configured for GDPR and COPPA compliance if your audience includes minors.

Prompt Engineering Principles for Ethical Review

AI's power comes with a responsibility to use it wisely. The tools we've discussed can accelerate every phase of game development, but they can also inadvertently introduce biases, ethical pitfalls, or legal risks. This section isn't about halting your creative process; it's about building a final, proactive layer of review to ensure your game is fair, transparent, and respectful to both your players and your creative peers. By using AI as an ethical watchdog, you can catch potential problems before they become public issues.

Prompt for Bias Audits: Use prompts to actively search for and correct biases in AI-generated content.

For example, *"Analyze the following NPC dialogue for any cultural, gender, or racial stereotypes. Suggest alternative phrasing that is more respectful and inclusive."*

Prompt for Player-Facing Disclosures: Use AI to draft clear, concise, and honest disclosures about AI use in the game.

For example, *"Draft a short, player-facing message for our game's start screen that explains our use of AI for dynamic NPC dialogue and content generation."*

Prompt for Risk Mitigation: Use AI to brainstorm solutions for ethical risks.

For example, *"We are using AI-generated voice lines. What are three potential legal or ethical risks, and how can we mitigate them?"*

AI Should Not Replace What You Love

We have mentioned this before: AI is a tool to help you achieve your goals. Don't let it control the development process. If you became a game developer because you love drawing concept art, designing levels, composing music, or writing dialogue, don't let AI take that away from you.

AI is best used to:

- Automate the tasks you *don't* enjoy.

- Speed up repetitive or technical work so you can focus on the creative elements that matter most to you.

- Provide inspiration when you're stuck, not dictate your artistic direction.

In *Lantern Paths*, we used AI to quickly create placeholder props, generate NPC dialogue drafts, and experiment with new music styles. But every piece of final art, every puzzle design, and every emotional beat in the story was guided by human vision and love for the craft.

A game is more than the sum of its assets—it's a reflection of its creators. If you hand over the parts you care about most, you risk losing the soul of your game.

AI Ethics for Small Teams: A Checklist

Before shipping your game, ask:

- Have we disclosed AI-generated content where relevant?

- Do we have the right to all training data?

- Have we reviewed AI outputs for harmful bias or misinformation?

- Are we complying with privacy laws in all markets we serve?

- Are we still directly doing the parts of development we're most passionate about?

- **Ethical Pitfall: The Creative Ownership Trap**
  This refers to the legal and philosophical pitfall of a developer losing track of what they created versus what the AI generated. When a developer builds a game entirely on AI-generated assets and code, do they still have a unique creative vision? This can lead to issues with copyright, but more importantly, it can lead to a developer feeling disconnected from their own work.

The Future of AI Ethics in Games

As AI becomes a standard part of development, ethical considerations will only grow in importance. What feels optional today, such as documenting AI-assisted code or being transparent about generated assets, will soon become expectations from players, publishers, and even regulators.

Studios that commit early to responsible AI practices will gain a competitive advantage. Ethical development isn't just about avoiding legal or reputational risks; it's about positioning your studio as a place where innovation and integrity go hand in hand. Developers increasingly want to work for teams that value fairness and sustainability, and players are more loyal to games they can trust. By showing that your AI use is thoughtful, transparent, and respectful, you'll build stronger communities around your projects.

Looking forward, we may see industry-wide standards emerge. Just as accessibility guidelines and content ratings became normal, AI ethics may be codified into frameworks that studios adopt by default. Those who already practice ethical AI development will adapt easily, while those who cut corners may struggle to catch up.

In short, the studios that thrive will be the ones that embrace AI as a tool for empowerment while keeping their commitment to creativity, fairness, and human connection at the center of their games.

Final Thoughts

Our hope is that you have found AI for Game Developers a useful guide as we navigate a rapidly changing world. The journey from a blank page to a finished game is an incredible one. We've explored how to use AI to find your core concept, build the foundation of your GDD, and turn greybox levels into polished, playable spaces. We've seen how these tools can help small teams achieve what once required the resources of a full studio, from generating assets in hours rather than months to creating a detailed marketing calendar for launch. This guide is about more than learning new software; it's about a new way of thinking—using AI to accelerate your creativity and amplify your strengths so you can achieve your vision.

Throughout this book, we have maintained that AI is a tool. The ethics and the joy come from how we choose to use it. The key to successful AI-assisted development is to be a skilled, intentional creator who never surrenders their creative control. You must still be the one to curate and refine the AI's output, ensuring your creative fingerprint is on every detail. The true purpose of AI is to handle the tedious, repetitive work, allowing you to focus on the parts of game development you love most. Whether that is designing levels, writing dialogue, or fine-tuning combat, your passion is the most important engine in the process. A game is more than the sum of its parts; it is a reflection of its creators. AI is just here to give your creativity more horsepower.

May your passion for games never fade, and your projects always find their audience.

Dr. Brian G. Burton

Bonus Chapter: Marketing and Promoting Your Indie Game

Indie marketing is less about budgets and more about bravery, the bravery to share your work before it feels perfect.

How This Chapter Differs from Chapter 11

In Chapter 11, we focused on marketing your specific game, building a tailored plan around your project, your team, and your immediate goals. That chapter served as a roadmap for your unique launch.

This bonus chapter steps back to look at the bigger picture of indie game marketing. Here, we'll explore timeless principles, proven strategies, and modern tools that can help you not only market your current game but also build momentum for every project you release in the future. If Chapter 11 was the map, this chapter is the compass: a set of guiding truths to keep you on course no matter where your creative journey leads.

Introduction: Why Marketing Matters as Much as Development

A common myth among indie developers goes like this: *"If the game is good enough, players will find it."*

The reality? Many excellent games go unnoticed because they never reach the right audience. In today's crowded marketplaces — Steam, Epic, Nintendo, Xbox, and mobile stores — visibility is just as important as quality.

Team Cherry's *Hollow Knight* proves the point. Three developers, starting with a modest Kickstarter in 2014, grew their game into one of the most beloved indies of all time. Their follow-up, *Silksong*, has built years of anticipation. This didn't happen by accident. It was the result of consistent storytelling, community-building, and marketing that made players feel part of something bigger.

Marketing isn't separate from development; it's part of faithfully bringing your game into the world.

Marketing is an act of stewardship; sharing what you've created so the people who need it can find it.

The Indie Advantage

Indie teams often see themselves at a disadvantage: no big budgets, no PR firms, no flashy campaigns. But being small comes with surprising strengths:

- **Agility:** You can pivot quickly, experiment with content, and respond directly to players.
- **Authenticity:** Players value real voices over corporate polish.
- **Storytelling:** Your journey as a small team is often as compelling as the game itself.

***Players don't just want to play your game
they want to join your journey.***

Understanding Your Ideal Player

Marketing always begins with people, not platforms. Before you think about trailers, posts, or ads, you need to know who you're talking to. We call this your *ideal player*.

Your ideal player isn't "everyone." If you try to make a game for everyone, you'll end up connecting with no one. Instead, think about the *specific kind of person* who will love your game most.

Step 1: Picture One Player, Not a Crowd

Imagine one person sitting down to play your game. Who are they?

- What other games do they already play?
- How do they spend their free time?
- What excites them about games?
- What frustrates them about games?

Step 2: Think in Plain Language

Don't overcomplicate this. You don't need a spreadsheet or a marketing degree. Just answer in everyday words:

- *"She's a college student who loves story-driven RPGs and spends hours on Twitch."*
- *"He's a teenager who plays cozy farming sims on his Switch and likes to relax after school."*

Step 3: Find Their Wants and Needs

Now ask: *what does this player want most?* Do they want a challenge? Relaxation? A sense of mastery? A social community?

And just as important: *what frustrates them?* Maybe games waste their time with filler. Maybe they're tired of paywalls. Maybe they want something fresh in a genre they love.

Step 4: Create a Player Promise

Once you know their wants and frustrations, you can write a simple "player promise." This is a short statement that explains why your game is perfect for them.

Here's a formula you can use:

Are you [type of player] who [frustration/problem], but you really want [desired outcome]? Our game helps [player type] achieve [specific outcome] without [common objection].

Example 1: Hollow Knight
"Are you a Metroidvania fan who loves exploration but hates endless grinding? *Hollow Knight* delivers rich worlds to explore and meaningful challenges without filler."

Example 2: Stardew Valley
"Are you a farming sim fan who loves cozy worlds but is tired of shallow mechanics? *Stardew Valley* offers depth, relationships, and seasons of discovery without microtransactions."

The clearer you are about who you serve, the easier it becomes to invite them into your vision.

Crafting a Clear Message

Once you know who your ideal player is, the next step is to decide what you want to say to them. This is your *message*.

A clear message is not a list of features. Players don't fall in love with "10 weapons, 4 levels, and customizable hats." They fall in love with the *experience* your game promises them.

Think of your message as a story you're telling:
- What problem does your game solve for players?
- Why does that problem matter?
- What could life look like if it were solved?
- How does your game deliver that better future?

This simple four-part flow gives your words direction.

Step 1: Identify the Problem

What frustrates your players about the games they already play?
- Too repetitive?
- Too expensive?
- Too shallow?

Step 2: Amplify Why It Matters

Show them why this problem is worth fixing.
- *"Every new release feels like a copy-paste."*
- *"You're paying more, but getting less."*

Step 3: Cast the Vision

Paint a picture of what it could be like if that problem didn't exist.
- *"Imagine a world where every corner is handcrafted and alive."*
- *"Imagine a game that relaxes you instead of stressing you out."*

Step 4: Present the Solution

Now show how your game delivers that vision.
- *"That's exactly the world we've built for you."*
- *"That's why our game gives you deep systems with no paywalls."*

Real-World Examples

Hollow Knight
- Problem: *"You love exploration games, but they feel repetitive."*
- Amplify: *"Every new release feels like a copy-paste."*
- Vision: *"Imagine a handcrafted world where every corner feels alive."*
- Solution: *"That's the experience Hollow Knight delivers."*

Celeste
- Problem: *"Platformers can feel shallow once you master them."*
- Amplify: *"You want challenge, but you also want meaning."*

- Vision: *"Imagine a game that pushes your skills while telling a heartfelt story."*
- Solution: *"Celeste combines precision platforming with an emotional journey."*

Try It Yourself: Your Message in 4 Steps

Fill in the blanks for your own game:

1. **Problem:** "Players love ___, but they're frustrated because ___."
2. **Amplify:** "That's a problem because ___."
3. **Vision:** "Imagine if ___."
4. **Solution:** "That's what our game delivers—___."

Your message is more than words; it's the lantern that shows players why your game matters.

Creating a Compelling Storefront

Your storefront (Steam page, itch.io page, mobile app listing, or console store listing) is often the first impression players get of your game. In many cases, it's the deciding factor between someone clicking "buy" or moving on.

Think of your storefront as your game's digital front door. If the door looks boring or confusing, players won't step inside. If it looks inviting and clear, they'll want to explore.

Step 1: Lead with a Strong Trailer

Your trailer is your handshake. Players usually watch only the first 10–15 seconds before deciding if they care.

- Show actual gameplay right away.
- Keep it short (60–90 seconds).
- End with a clear call to action: "Wishlist now" or "Available on Steam."

Step 2: Use Screenshots that Tell a Story

Screenshots are like the pictures on the back of a book. They should show players *why this game is exciting*.

- Show variety: combat, exploration, cozy moments, dramatic scenes.
- Avoid only showing menus, maps, or inventory screens.
- Make each screenshot an invitation to imagine playing.

Step 3: Write a Description That Speaks to Players

Don't just list features ("10 levels, 3 bosses"). Instead, describe the *experience*:

- *"Explore a mysterious underground world full of secrets."*
- *"Build the farm of your dreams while making friends in town."*
- *"Test your reflexes in a fast-paced arena with endless challenges."*

Your description should make players think: *"That's the kind of adventure I want."*

Step 4: Optimize for Search (Keywords)

On Steam, mobile, and console stores, **keywords** help players find your game. Over half of mobile games are discovered through **search**, and Steam is moving in the same direction.

- Use the words your ideal players actually type: "Metroidvania," "roguelike," "cozy farming sim," "souls-like."
- Sprinkle those words naturally into your description.
- Avoid vague or generic terms like "fun" or "exciting."

Real-World Examples

Hollow Knight (Steam page)

- Trailer: Starts with haunting visuals and immediately jumps into gameplay.
- Screenshots: Mix of exploration, combat, and boss fights.
- Description: Emphasizes "vast, interconnected world" and "epic battles."
- Keywords: "Metroidvania," "exploration," "challenging."

Stardew Valley (Steam page)

- Trailer: Shows farming, relationships, and town life in the first 15 seconds.
- Screenshots: Farming, fishing, building, friendships.
- Description: "Create the farm of your dreams" connects emotionally to players.
- Keywords: "Farming sim," "RPG," "cozy."

Try It Yourself: Your Storefront Blueprint

Fill in the blanks for your own game:

1. **Trailer:** "In the first 15 seconds, I will show ___."
2. **Screenshots:** "I will include images of ___, ___, and ___."
3. **Description:** "My one-sentence hook for players is: ___."
4. **Keywords:** "The top 3 words my players search for are ___, ___, and ___."

> *Your storefront is your handshake with the world.*
> *Make it clear, confident, and memorable.*

Proof and Process: Show, Don't Just Tell

When you're trying to convince players to try your game, your words aren't enough. Players believe other players more than they believe you. That's why proof is so important. Proof shows that your game is real, fun, and worth their time.

Think of proof as evidence. Just like a restaurant shows happy diners in photos or a band shares reviews from fans, you need to show that real people are already enjoying your game.

Step 1: Collect Player Feedback Early

Don't wait until launch to hear from players. Share early builds with friends, testers, or a small community. Gather their comments and highlight the best ones.

- *"The combat feels tight and rewarding."*
- *"I couldn't stop exploring once I started."*

Step 2: Share Visual Proof

Screenshots and short clips of people playing your game are powerful. If someone laughs, gasps, or says "one more try!" on stream, that's proof your game works.

- Clip highlights from playtests or Twitch streams.
- Share fan art or screenshots that players post.

Step 3: Amplify Influencer and Tester Quotes

Even small streamers and micro-influencers can help. A single positive line from them—*"This boss fight was brutal but fair!"*—is more convincing than paragraphs of marketing copy.

Step 4: Highlight Your Community

Players love seeing that they're joining something bigger. Share contests, fan art, or even funny bugs your community discovers. It shows your game is alive.

Real-World Examples

Among Us (early on)
The game was small until streamers started showing proof of hilarious social interactions. Those videos became the marketing.

Celeste
Quotes from reviewers and players about how the story "stuck with them" were highlighted on the storefront and in trailers. Proof of emotional impact drove more sales than features ever could.

Try It Yourself: Your Proof Plan

Fill in the blanks for your own game:

1. **Feedback:** "I will gather quotes from ___ and share them on ___."
2. **Visual Proof:** "I will post clips/screenshots of players reacting to ___."
3. **Influencers:** "I will reach out to ___ small creators and highlight their comments."
4. **Community:** "I will showcase fan art, speedruns, or funny moments like ___."

> *Player wish lists are a future player raising their hand and saying, 'I want to see your world.'*

Timing and Launch Strategy

Launching your game is like planning a big event. You wouldn't throw a wedding or a concert without picking the right date, inviting people ahead of time, and making sure everything is ready. The same is true for your game launch.

A good launch strategy isn't just about one day; it's about the build-up, the big moment, and what happens right after.

Step 1: Pick the Right Time

Don't launch your indie game the same week as a massive AAA title. If a big release like *Elden Ring* or *Zelda* drops, the gaming world will be focused there. Choose a quieter window when your game has a better chance of standing out.

Step 2: Build Anticipation with Demos

Players love to try before they buy.

- Release a demo during Steam Next Fest or on itch.io.
- A demo gives players a taste, builds wishlists, and provides feedback to improve the game before launch.

Step 3: Soft Launch for Testing

A soft launch means releasing to a smaller group before going worldwide.

- You might release only in certain regions (like Canada or Australia for mobile games).
- Or you might invite a limited number of testers through Discord.
- Use this stage to polish onboarding, fix bugs, and see how new players respond.

Step 4: Plan the Global Launch

Your global launch is your "grand opening." Prepare ahead of time:

- **Press kit:** Screenshots, trailers, descriptions, and logos.
- **Outreach:** Contact influencers, streamers, and journalists.
- **Community events:** Host a stream, AMA (Ask Me Anything), or Discord launch party.

Step 5: Keep the Momentum Going

A launch isn't over on day one.

- Post updates on how the launch went.

- Thank your community publicly.
- Share reviews, fan art, or clips from new players.

Real-World Examples

Hollow Knight
Team Cherry kept anticipation alive by consistently updating their Kickstarter backers and later expanding with new free content, extending the "launch moment" over time.

Stardew Valley
The developer released small updates after launch, which not only fixed issues but created new reasons for players to return and spread the word.

Try It Yourself: Your Launch Plan

Fill in the blanks for your own game:

1. **Pick the Time:** "I will avoid launching near ___ and aim for ___."
2. **Demo:** "I will release a demo on ___ to gather feedback and wishlists."
3. **Soft Launch:** "I will test the game first with ___ players/community."
4. **Global Launch:** "I will prepare a press kit with ___, reach out to ___, and host ___."
5. **Momentum:** "After launch, I will thank my players by ___."

Pro Tip: Soft Launching
Test in smaller markets, refine onboarding, fix issues, and build buzz *before* global release.

Don't rush the unveiling. Test, refine,
and then reveal with confidence.

A Simple Repeatable Strategy

Marketing feels overwhelming when you try to be everywhere, do everything, and chase every trend. The secret is not doing *more*, but doing what you can do **consistently**.

Think of marketing like exercise: one workout won't get you in shape, but steady, repeatable habits will. The same goes for reaching players—small, regular actions build momentum over time.

Step 1: Pick One Main Platform

Don't try to master Twitch, YouTube, TikTok, Twitter, Reddit, and Instagram all at once. Pick the platform where your *ideal players* already spend their time.

- Making short funny clips? TikTok or YouTube Shorts.
- Sharing devlogs and art? Twitter/X or Reddit.
- Building community? Discord.

Focus there first. You can always expand later.

Step 2: Choose a Simple Rhythm

Decide how often you can realistically post.

- **Daily**: Quick updates, behind-the-scenes images, or fun questions on Discord.
- **Weekly**: A devlog video, blog post, or short trailer clip.
- **Monthly**: A bigger event like a demo update, livestream, or community challenge.

The exact schedule isn't as important as keeping it steady.

Step 3: Reuse and Repurpose

Don't create everything from scratch. If you record a short devlog, turn it into:

- A blog post.
- A Twitter thread.
- A Discord announcement.
- A short clip for TikTok.

One piece of content can go a long way.

Step 4: Engage, Don't Just Broadcast

Marketing isn't shouting through a megaphone. It's more like sitting at a campfire.

- Reply to comments.
- Ask your community questions.
- Share fan art or memes they make.

When players feel heard, they're more likely to stick around.

Real-World Examples

ConcernedApe (Stardew Valley)
He didn't post daily, but when he did, it was consistent: dev updates, patch notes, and sneak peeks. That steady rhythm kept players engaged for years.

Inkle Studios (80 Days, Heaven's Vault)
They reused story snippets, art, and behind-the-scenes sketches across platforms, making a little content go far.

Try It Yourself: Your Repeatable Strategy

Fill in the blanks for your own game:

1. **Main Platform:** "I will focus on ___ first."
2. **Daily/Weekly Rhythm:** "I will share ___ every ___."
3. **Monthly Event:** "Once a month, I will ___."
4. **Repurposing:** "From one devlog, I will also create ___, ___, and ___."
5. **Engagement:** "Each week, I will reply to ___ comments and highlight ___ pieces of community content."

Consistency creates trust. Trust creates community. Community creates momentum.

Automating Without Losing Authenticity

When indie developers hear the word *automation*, they sometimes think it means "robots spamming posts" or "losing the personal touch." But automation doesn't mean being fake; it means setting up simple systems so you don't burn out.

Think of automation like setting a coffee maker the night before. You still enjoy the coffee, but the machine helped you save time in the morning. Marketing works the same way.

Automation lets you spend less time on repetitive tasks and more time on what matters: building your game and connecting with players.

Step 1: Schedule Your Content

Instead of rushing to post something new every day, plan ahead.

- Use free tools (like Buffer, Hootsuite, or built-in scheduling on Twitter/X, YouTube, and TikTok).
- Write a week's worth of posts in one sitting.
- Schedule them to go live automatically.

This keeps your game visible, even when you're focused on development.

Step 2: Automate Email Updates

Your mailing list is a direct line to players. You can write one email and have it automatically sent to:

- New sign-ups ("Thanks for joining—here's a free wallpaper").
- Players who wishlist your game.
- Fans who want launch-day reminders.

Automated email means players stay informed without you manually sending each update.

Step 3: Create a Press Kit Hub

Journalists, influencers, and streamers often want screenshots, trailers, or logos. Instead of answering the same questions 20 times, create a press kit page on your site with everything they need.

It's automated in the sense that once it's there, it works for you—no extra effort required.

Step 4: Protect Your Authentic Voice

Automation is a helper, not a replacement.

- Always write posts in your own style.
- Balance scheduled content with live interactions (like replying to comments or chatting on Discord).
- Use automation for consistency, and real conversations for connection.

Real-World Examples

Team Cherry (Hollow Knight)
Their Kickstarter updates were scheduled but written with a personal tone. Players felt connected because the voice was authentic, even if the timing was automated.

Solo Devs on Twitter
Many indie devs use scheduled devlogs, but still hop into the replies personally. That mix of automation + live presence builds trust.

Try It Yourself: Your Automation Plan

Fill in the blanks for your own game:

1. **Scheduling:** "I will schedule posts on ___ once per week so they appear every ___."
2. **Email:** "When someone signs up for my mailing list, they will automatically receive ___."
3. **Press Kit:** "I will create a press kit with ___, ___, and ___ to save time answering requests."
4. **Balance:** "I will combine automated posts with live check-ins by ___."

Automation is not replacing your voice; it's freeing it to speak where it counts most.

Sustaining Momentum Post-Launch

Many indie developers think the story ends on launch day. But in reality, launch day is just the *start* of a new chapter. The days, weeks, and months after release determine whether your game quietly fades or continues to grow.

The good news? Post-launch marketing doesn't have to be overwhelming. With a few simple steps, you can keep your community engaged, attract new players, and build a lasting reputation.

Step 1: Release Meaningful Updates

Updates show players you care. They don't all have to be huge expansions—sometimes small changes make a big impact.

- Fix common bugs quickly.
- Add small quality-of-life improvements.
- Release free extras like new skins or levels.

Players notice when a developer keeps improving the game.

Step 2: Host Community Events

Invite players to participate in your world.

- Run speedrun contests.
- Share fan art competitions.
- Create polls and let the community vote on small features.

This transforms players from customers into collaborators.

Step 3: Plan Bigger Content Drops

When you're ready, add expansions or new modes. These are opportunities to reignite press coverage and influencer interest.

- *Hollow Knight* extended its life with free content packs.
- *Stardew Valley* continued to grow with major updates that brought players back years later.

Step 4: Make Player Support Part of Marketing

Every email, comment, or bug report is a chance to build trust. Respond kindly and promptly. Show your fixes publicly so players see you're listening.

- Fast support builds loyalty.
- Good support leads to positive reviews.
- Reviews boost visibility in storefronts.

Support isn't just customer service—it's marketing through *trust*.

Real-World Examples

Among Us
The devs could have moved on after launch, but consistent updates, maps, and community-driven tweaks kept the game alive long enough for streamers to discover it and make it a worldwide hit.

No Man's Sky
A rocky launch turned into a redemption story because the developers kept releasing meaningful updates and supporting the community long after the world stopped paying attention.

Try It Yourself: Your Post-Launch Plan

Fill in the blanks for your own game:

1. **Updates:** "In the first 3 months, I will release updates that fix ___ and add ___."
2. **Events:** "I will host a community event such as ___ to encourage players to share their experiences."
3. **Expansions:** "At ___ months after launch, I will plan a bigger content drop like ___."
4. **Support:** "I will respond to player feedback within ___ days and share improvements through ___."

Pro Tip: Player Support
Every support interaction is marketing. Fast, helpful responses create loyalty and generate positive reviews.

Every update and every reply is a promise kept, and promises kept build lasting loyalty.

Common Pitfalls to Avoid

Most indie games don't fail because of poor design. They fail because players never hear about them. Avoiding common marketing mistakes can save you months of frustration and give your game a real chance to succeed.

Think of pitfalls as potholes in the road: they don't stop your journey, but they can slow you down or wreck your momentum if you're not careful.

Pitfall 1: No Visibility Plan

Some developers launch without building a community, collecting wishlists, or thinking about keywords. The result? Even if the game is great, it's invisible.

- Solution: Start early. Collect wishlists. Share devlogs. Optimize your storefront with keywords.

Pitfall 2: Bad Timing

Releasing against a blockbuster game can bury your indie project. AAA studios have massive marketing budgets—you can't out-shout them.

- Solution: Watch the release calendar. Aim for a quieter window where your game has breathing room.

Pitfall 3: Neglecting Player Support

Ignoring emails, reviews, or bug reports frustrates players and can lead to negative word-of-mouth.

- Solution: Treat every player interaction as marketing. A fast, kind response creates loyalty and positive reviews.

Pitfall 4: Doing Too Much, Too Soon

Trying to be on every platform, run every event, and post daily on five social channels is a recipe for burnout.

- Solution: Focus. Pick one main platform, one simple strategy, and grow from there.

Real-World Examples

Pitfall: No Visibility
Thousands of Steam games quietly release each year with no wishlists or community. Most disappear within days.

Pitfall: Bad Timing
Many indies that launched the same week as *Elden Ring* or *Tears of the Kingdom* saw their sales crushed—not because they were bad, but because they were drowned out.

Pitfall: Neglecting Support
Even big studios with poor support see review scores plummet. For indies, support is often the difference between a "Mostly Positive" and a "Mixed" rating.

Try It Yourself: Pitfall Checkup

Answer these questions before launch:

1. **Visibility:** "Have I collected at least ___ wishlists and posted at least ___ devlogs?"
2. **Timing:** "Am I avoiding the launch window of ___ big games?"
3. **Support:** "How will I reply to bug reports or reviews within ___ days?"
4. **Focus:** "Which one platform or community will I prioritize first?"

Final Checklist

Before Launch

- Define your *ideal player*.
- Write your "player promise."
- Craft your problem → amplify → vision → solution message.
- Launch your Steam page early and collect wish lists.
- Build scalable assets (trailer, press kit, devlog).
- Optimize keywords and screenshots.

During Launch

- Share proof: play-tester reviews, fan art, influencer clips.
- Repurpose your trailer everywhere.
- Run a launch-day community event.
- Automate posts and press outreach.
- Use soft launches to refine before going global.

After Launch

- Track KPIs: wish lists, conversions, retention.
- Release meaningful updates.
- Provide fast, supportive player responses.
- Showcase community content.
- Plan expansions or bundles.

Conclusion

Marketing your indie game isn't about being everywhere or spending big. It's about consistently showing up, speaking clearly, and having the courage to share your vision even when it doesn't feel perfect yet.

Every post is a seed. Every connection is a bridge. Every update is a promise kept.

Keep showing up. Keep speaking clearly. Keep shining your light. The players who need your game are already waiting to find you.

Appendix A – AI Tool Quick Reference

AI Tools Featured in This Book

This appendix brings together all of the AI tools mentioned across chapters. Each is grouped by the phase of development it supports, with a brief description and link.

This is not an exhaustive list. New AI tools are appearing all the time. **If you know of a tool we should have included, or if you've found one that helped you in a way we didn't cover, please let us know**. Your feedback will help shape the next edition of this guide.

Pre-Production & Concept Design

- **Artbreeder** – Collaborative AI for generating and "breeding" portraits, landscapes, and concept visuals.
  https://www.artbreeder.com

- **ChatGPT / Claude / Gemini / Grok** – Large language models for ideation, brainstorming mechanics, drafting GDD sections, and narrative assistance.
  https://chat.openai.com | https://claude.ai | https://gemini.google.com | https://grok.x.ai

- **Ludo.ai** – Specialized for game ideation, trend analysis, concept scoring, and GDD templates.
  https://ludo.ai

- **Perchance Generators** – Randomized generators for game twists, names, and scenario mashups.
https://perchance.org

- **Sudowrite** – AI writing assistant for creative text, narrative ideas, dialogue, and character personalities.
https://www.sudowrite.com

3D Modeling & Asset Creation

- **Adobe Firefly** – AI for generating textures, vector art, and edits directly in Adobe's suite.
https://firefly.adobe.com

- **Hitem3D.ai** – Creates optimized 3D assets ready for Unity/Unreal.
https://hitem3d.ai

- **Kaedim** – Converts 2D concept art into 3D models.
https://www.kaedim3d.com

- **Layer** – Cross-platform AI for both 2D and 3D game assets.
https://layer.ai

- **Meshroom** – Open-source photogrammetry for turning photos into 3D models.
https://alicevision.org/#meshroom

- **Meshy.ai** – Generates 3D models and textures from text or images.

https://www.meshy.ai

- **Sloyd 2.0** – Browser-based Text-to-3D and Image-to-3D generator, optimized for indies.
  https://www.sloyd.ai

Level Design & Prototyping

- **Blockade Labs** – Creates 360° skyboxes and immersive environments.
  https://www.blockadelabs.com

- **Buildbox 4** – Rapid prototyping and AI-assisted level creation.
  https://www.buildbox.com

- **Intangible** – Spatial AI for sketching interactive 3D layouts in-browser.
  https://www.intangible.ai

- **Ludo.ai Playable Generator** – Generates basic playable prototypes from prompts.
  https://ludo.ai

- **Masterpiece Studio** – AI modeling and rigging tools for placeholder assets.
  https://www.masterpiecestudio.com

- **Speedtree** – Industry tool for procedural trees and foliage.

https://speedtree.com

- **Superhive** – Blender add-ons for procedural content generation.
https://superhivemarket.com

- **World Creator** – Real-time terrain generation for rapid iteration.
https://www.world-creator.com

- **World Machine** – Procedural terrain generator with erosion simulation.
https://www.world-machine.com

Writing, Dialogue & Worldbuilding

- **Convai** – Real-time, voice-driven NPC interactions.
https://convai.com

- **Inworld AI** – Creates NPCs with persistent personalities, memories, and emotions.
https://inworld.ai

- **Notion AI / Obsidian** – AI-enhanced organization for lore bibles and searchable notes.
https://www.notion.so/product/ai | https://obsidian.md

Character Creation, Animation & Voice

- **Audio2Face** – NVIDIA tool for automatic lip-sync animation.
  https://build.nvidia.com/nvidia/audio2face-3d

- **Cascadeur** – AI-assisted physics-based animation.
  https://cascadeur.com

- **DeepMotion** – Motion capture from video.
  https://www.deepmotion.com

- **ElevenLabs.io** – Natural, emotive AI voice synthesis.
  https://elevenlabs.io

- **Inworld AI** – Personality-driven NPCs with emotions, memories, and goals.
  https://inworld.ai

- **Radical Motion** – Real-time body tracking.
  https://radicalmotion.com

- **Ready Player Me** – Cross-platform avatar creation for players.
  https://readyplayer.me

- **Reallusion Character Creator** – AI-assisted morphing, rigging, and stylized characters.
  https://www.reallusion.com/character-creator/

- **Unreal Metahuman** – Realistic, fully rigged human avatars.

https://www.unrealengine.com/en-US/metahuman

- **Uthana** – Generative motion from text or video input, trained on mocap data.
  https://uthana.com

Audio: Music & Sound

- **Aiva** – Orchestral and adaptive soundtrack generation.
  https://www.aiva.ai

- **AudioStrip** – Isolates music from dialogue/effects.
  https://audiostrip.co.uk

- **Boomy** – Fast and simple AI music creation.
  https://boomy.com

- **Eleven Labs** - Provides music and sound effects.
  https://elevenlabs.io

- **LALAL.AI** – Splits tracks into stems for remixing.
  https://www.lalal.ai

- **Sononym** – AI-assisted sound organization and sample discovery.
  https://www.sononym.net

- **Suno** – AI music generation with lyrics and genre control.
  https://suno.ai

Programming & Gameplay Systems

- **Aura** – AI visual scripting tool for Unity.
  https://aura.one

- **Claude (Anthropic)** – Long-context LLM for explaining/debugging code.
  https://claude.ai

- **Cursor** – AI-powered IDE for iteration and debugging.
  https://cursor.com

- **GitHub Copilot** – Real-time AI code completions.
  https://github.com/features/copilot

- **Ludus Engine** – Generates Unreal Engine scripts and Blueprints.
  https://ludusengine.com/

- **Machinations** – AI analysis and balancing of game economies.
  https://machinations.io

- **Ponicode** – Automatic unit test generation.
  https://www.ponicode.com

- **Replit Ghostwriter** – Browser-based AI coding.
  https://replit.com/ghostwriter

- **Vibe Coding Tools to Watch**

- **Aider** – https://aider.chat

- **Eve Programming Environment** – https://witheve.com

- **Qodo** – https://www.qodo.ai

- **Tabnine** – https://www.tabnine.com

Testing & Analytics

- **GameAnalytics AI** – Player behavior and retention analytics.
  https://gameanalytics.com

- **Helicone** – Monitors AI usage and backend performance.
  https://helicone.ai

- **Keen.io** – Player data analytics with ML insights.
  https://keen.io

- **modl.ai** – Playtesting bots and balancing analytics.
  https://modl.ai

- **nunu.ai** – Automated QA for bug detection.
  https://nunu.ai

- **Testify AI** – Simulates human playthroughs for QA.
  https://tudip.com/products/testify-ai/

Marketing & Community Engagement

- **ChatGPT / Gemini** – Store copy, press releases, and social media content.
  https://chat.openai.com | https://gemini.google.com

- **Midjourney / DALL·E / Stable Diffusion / Firefly** – Key art and marketing visuals.
  https://www.midjourney.com | https://openai.com/dall-e | https://stability.ai/stable-diffusion | https://firefly.adobe.com

- **Pika Labs** – AI-assisted text-to-video and short promotional clips.
  https://pika.art

- **Runway** – AI-assisted video editing and trailer production.
  https://runwayml.com

- **Synthesia** – AI avatar-based video announcements.
  https://www.synthesia.io

How to Use This Appendix

- Test each tool in a low-risk environment before integrating it into your production pipeline.

- Track license terms, as AI usage rights can change over time.

Appendix B – Game Design Document Template

AI-Integrated Indie Game GDD

A Game Design Document (GDD) is your game's blueprint—a living reference that keeps everyone on your team (even if it's just you) working toward the same vision. It describes what your game is, how it plays, how it looks and sounds, and how it will come together. While big studios often produce massive GDDs, an indie-friendly GDD can be short, visual, and practical. This template is designed to be flexible and approachable, with optional "AI Notes" so you can track where artificial intelligence tools are helping your workflow. Even if you've never made one before, filling this out will save you time, reduce rework, and make it easier to communicate your ideas to collaborators, investors, and players.

---

1. Game Overview

Game Title:
Tagline: *(1 sentence pitch)*
Genre:
Platform(s):
Target Audience:
Estimated Release Date:

High Concept:
(Brief summary of the game—one paragraph describing what makes it unique)

AI Notes:
(Optional – list AI tools used in conceptualization; e.g., Ludo.ai for ideation, Midjourney for mood boards)

---

2. Core Gameplay

Player Goals:
Primary Mechanics:
Secondary Mechanics:
Progression System:

AI Notes:
(E.g., Using Ludus.ai for prototype layouts, Scenario.gg for procedural props)

---

3. Story & Narrative

Premise:
Setting:
Main Conflict:
Themes:
Characters:

- Name, role, backstory, personality traits

- AI-assisted dialogue tools used (Inworld AI, Convai)

Story Flow:
(High-level outline of acts, chapters, or missions)

---

4. Art & Visual Style

Style Reference: *(e.g., 2D pixel art, stylized 3D, photorealistic)*
Color Palette:
UI/UX Style Guide:

AI Notes:
(E.g., Midjourney for concept art, Scenario.gg for asset variations)

---

5. Audio

Music Direction: *(style, instruments, mood)*
SFX Style:
Voice Acting: *(casting notes, tone)*

AI Notes:
(E.g., Aiva for adaptive music, ElevenLabs.io for prototype voices)

---

6. Level & Environment Design

World Layout: *(maps, zones, regions)*
Key Locations:
Level Flow: *(player movement through spaces)*

AI Notes:
(E.g., Ludus.ai for layout drafting, Mystic prototype generator for playable mock-ups)

---

7. Technical Specifications

Engine: *(Unity, Unreal, Godot, etc.)*
Platform Requirements:
Multiplayer/Singleplayer:
Networking Needs:

AI Notes:
(E.g., AI-powered performance profiling, GameAnalytics integration)

---

8. Monetization Strategy *(If Applicable)*

Model: *(premium, free-to-play, DLC, cosmetics)*
In-Game Store: *(items, currencies)*
Ethical Considerations:

---

9. Marketing & Community Plan

Target Channels: *(Discord, Steam, social media)*
Core Messaging:
Launch Strategy:

AI Notes:
 (E.g., ChatGPT for press releases, Midjourney for social media teasers)

---

10. Production Timeline & Milestones

Milestone 1: Concept Approval – Date
Milestone 2: Prototype – Date
Milestone 3: Alpha – Date
Milestone 4: Beta – Date
Milestone 5: Release – Date

AI Notes:
 (E.g., using AI to rapidly create test assets and iterate faster between milestones)

---

11. Risks & Mitigation

Risk:
Impact:
Mitigation Strategy:

AI Notes:
(E.g., reviewing AI outputs for copyright, bias, and ethics concerns)

---

12. Appendices

- Concept Art Gallery *(AI-assisted and human-created)*

- Prototype Screenshots

- AI Tool Usage Log

Appendix C – Indie Game Business Plan Template

For Indie and Small Studio Projects

Introduction

A business plan might sound like something only large companies need, but it's just as valuable for small teams and solo developers, especially if you plan to fund your project, market it, or turn it into a sustainable business. This template will guide you through defining your game's market, setting realistic budgets, and planning your path to release. You don't have to be an accountant or MBA to use it, just fill it out honestly and keep it updated as your project evolves.

We've included "AI Notes" so you can track how AI tools save time and money, giving you a clearer picture of your true costs and potential profits. Think of this as a roadmap for turning your creative vision into a finished, profitable game.

---

1. Executive Summary

Game Title:
Tagline: *(1–2 sentences capturing the game's appeal)*
Studio Name:
Core Team Members:
Planned Release Date:
Funding Goal: *(if applicable)*

High-Level Pitch:
A short, engaging description of the game and why it will succeed. Include your target audience, unique selling points, and how AI integration enhances production.

---

2. Company Overview

Studio Mission Statement:
Founding Date:
Team Structure & Roles:
Experience & Track Record:

AI Notes:

- Highlight AI as a strategic advantage for speed, iteration, and scope management.

- Mention any AI partnerships or beta tool access.

---

3. Product Description

Genre & Core Features:
Target Platforms:
Key Innovations: *(both gameplay and technical)*

AI Notes:

- AI-powered NPCs (Convai, Inworld)

- AI-assisted art pipelines (Midjourney, Scenario.gg)

- Level prototyping (Ludus.ai, Mystic)

---

4. Market Analysis

Target Audience: *(age, location, gaming preferences)*
Market Size & Growth Trends:
Competitive Analysis: *(list similar games and explain your edge)*

AI Notes:

- Use AI market research tools to generate competitor data and discover underserved niches.

---

5. Marketing Strategy

Brand Positioning:
Channels: *(Steam, Epic Store, Itch.io, Kickstarter, social media, events)*
Content Plan: *(Devlogs, behind-the-scenes videos, influencer outreach)*
Community Building: *(Discord, beta testing groups, mailing list)*

AI Notes:

- AI content generation for social media (ChatGPT)

- AI-driven ad targeting suggestions

- AI-generated promotional images/videos

---

6. Development Plan

Project Phases: *(Pre-production, Production, Alpha, Beta, Release)*
Milestones:
Tools & Software: *(Game engine, asset creation tools, AI tools)*

AI Notes:

- Rapid prototyping with AI tools to shorten pre-production cycles.

- Placeholder asset generation to keep momentum.

---

7. Operations Plan

Team Roles & Responsibilities:
Workflow & Communication Tools: *(Trello, Slack, Notion, etc.)*
Outsourcing & Contractors: *(voice actors, localization, QA)*

AI Notes:

- Using AI to supplement contractor work and reduce outsourcing costs.

---

8. Financial Plan

Startup Costs: *(licenses, hardware, software, marketing, AI tool subscriptions)*
Revenue Streams: *(game sales, DLC, in-game cosmetics, Patreon)*
Pricing Strategy:
Break-even Analysis:

Projected Budget Table:

Category	Cost Estimate	AI Savings	Notes
Art Assets	$	$	AI-generated placeholders via Scenario.gg
Voice Acting	$	$	AI prototyping via ElevenLabs.io
Level Design	$	$	AI-assisted layouts via Ludus.ai

9. Risk Assessment

Market Risks: *(e.g., market saturation)*
Technical Risks: *(e.g., AI tool API changes)*
Legal Risks: *(e.g., copyright disputes)*

Mitigation Plan:

- Use only AI outputs with clear licensing.

- Have fallback workflows if AI tools become unavailable.

---

10. Appendices

- GDD (from Appendix B)

- Detailed Financial Projections

- Marketing Asset Samples

- AI Tool Usage Log

This plan is concise enough for Steam page preparation or Kickstarter pitches, but also scalable for grant applications and investor decks.

www.ingramcontent.com/pod-product-compliance
Lightning Source LLC
Chambersburg PA
CBHW071557210326
41597CB00019B/3280